REALLY SIMPLE

Clifford Burroughs

Published by New Generation Publishing in 2021

Copyright © Really Simple Future Limited 2021
www.really-simple.com

First Edition

The author asserts the moral right under the Copyright, Designs and Patents Act 1988 to be identified as the author of this work.

All Rights reserved. No part of this publication may be reproduced, stored in a retrieval system or transmitted, in any form or by any means without the prior consent of the author, nor be otherwise circulated in any form of binding or cover other than that which it is published and without a similar condition being imposed on the subsequent purchaser.

ISBN
    Paperback   978-1-80369-025-4
    Ebook   978-1-80369-026-1

www.newgeneration-publishing.com

New Generation Publishing

Content Overview

Preface

Introduction: It's Really Simple

1. Thinking: It's Complicated

2. Challenges: It's Counterintuitive

3. Benefits: It's Transformational

4. Model: It's Multidimensional

5. Mindset: It's Psychological

6. Method: It's Systematic

Tools: Support

Summary: It's Really Simple

PREFACE	1
INTRODUCTION: IT'S REALLY SIMPLE	7
1 THINKING: IT'S COMPLICATED	**11**
COMPLEXITY AND COMPLICATION	12
IT'S IN YOUR HEAD	15
TRIGGER UNHAPPY	24
IT'S MULTI-LEVEL	26
TIME: THE ETERNAL ENEMY	26
IT'S ABOUT PEOPLE	28
IT'S NOT A SILVER BULLET (OR A VALUE)	36
2 CHALLENGES: IT'S COUNTERINTUITIVE	**37**
BEWARE OF NONCHALANCE	39
INHERENT DIFFICULTY	45
NATURAL INSTINCT	53
BEING INSPIRED	58
3 BENEFITS: IT'S TRANSFORMATIONAL	**59**
BRANDS	59
COMMUNICATION	61
DECISIONS	65
PRODUCTS	67
SERVICES	69
PROCESSES	73
ACTIVITIES	76
4 MODEL: IT'S MULTIDIMENSIONAL	**78**
REDUCTIONS AND INCREASES	78
MEASURABLE (CHANGING) ATTRIBUTES	82

THE RED-CAR TEMPLATE	98
5 MINDSET: IT'S PSYCHOLOGICAL	**99**
I1: INTENTION (OR HIGHLY MOTIVATED)	99
I2: INTENSITY (OR SUSTAINED TENACITY)	103
I3: ITERATIONS (OR KEEP IMPROVING)	106
I4: INTELLIGENCE (OR GOOD THINKING)	108
I5: INVESTMENT (OR FINDING RESOURCES)	130
THE 5I'S TEMPLATE	132
6 METHOD - IT'S SYSTEMATIC	**134**
DASIA	135
DEFINE: CLARIFY THE CHALLENGE	137
ANALYSE: DEVELOP UNDERSTANDING	138
SYNTHESIS: DESIGNING RS SOLUTIONS	141
IMPLEMENT - CHANGE/CREATE	143
AFFIRM: REVIEW	145
DASIA TEMPLATE	146
TOOLS - SUPPORTING	**147**
TOOL 1: DF-PLAN	148
TOOL 2: S-OPRA	151
TOOL 3: I-CHARTER	159
TOOL 4: DAN-RICO?	163
TOOL 5: A-LIST	168
TOOL 6: EA-LIST	173
SUMMARY: IT'S REALLY SIMPLE	179
BIBLIOGRAPHY	181

Acknowledgements

Thanks to Amanda for her belief.

Thanks to Hollie for her encouragement.

Thanks to everyone else for your support.

Clifford Burroughs

Preface

Why I Wrote This Book

We all regularly experience complication. It can impact anyone. It is an issue for CEOs, managers, parents, workers, sports people, and kids growing up and getting ready for the world.

We waste huge amounts of time and energy on the complicated mechanics of working and living.

In contrast, simplicity has become a common aspiration in life. Business leaders, social commentators, my family, and friends all long for a world where "stuff could be simpler".

This frustration with the practicalities of work and life, combined with an innate instinct to make the world a slightly better place, was the motivation behind *Really Simple*. I started to look for best-practice solutions for the various efficiency and effectiveness problems I identified. These included a wide range of problems from email and document management, through process and systems design, to staying on top of the endless demands of our time.

The potential of simplicity as an enabler seemed to be a recurring theme for me, so I started proactively looking at almost everything from a simplicity perspective, capturing thoughts and examples that might potentially apply to work and life.

Although I thought that I knew what simplicity was, it became clear just how difficult it was to accurately articulate its definition, importance, and application in a straightforward way.

The irony was how difficult it was to explain simplicity simply!

I started to read, research, and observe everything I could sensibly find, including some now seminal and successful books. These included lean thinking, systems architecture, software design, process design, product/service design, organisational thinking, government policy, building design and personal productivity.

Another potential source of inspiration was the plethora of 'lifestyle' thinking, centred on simplicity. Typically, this focused on living a simple life, sometimes with a spiritual dimension and often with a cultural one. These include the topics of mindfulness, hygge, decluttering, paleo, veganism, yoga, and meditation, all of which are now in the mainstream (although not yet pervasive). I also reviewed the thought leadership at various consultancy firms, both big and boutique, and found mixed levels of commitment, understanding, and methodology.

Despite their undoubted value in helping me refine my ideas, nothing seemed to hit the mark addressing how to understand and apply simplicity in work and life.

The conclusion was that simplicity needed method!

In addition to extensive research, I tested out various methods and models in real life and business settings. I had also been exploring how to simplify executive life in big corporations and businesses. Executives were increasingly being asked to use a range of processes and tools that were unnecessarily complicated, although ironically those same executives were often the ones adding the complication.

I then faced a very specific business challenge of 'carving' out an integrated business (from a much larger one), to create a standalone business with a very different operating model. This is when I started to formally analyse, synthesise, and advocate the various aspects of simplicity and how to apply this thinking to real-world situations.

The end results transformed the way the business worked and tackled some big problems. We did this for a fraction of the cost of some of the more traditional solutions. We laid foundations that could help mitigate the increasing complication the organisation was facing and, more importantly, we could leverage new opportunities enabled by some of the macro trends in technology and supply chains, combined with an increase in simplicity expectations.

Along the way, I also started to think about my own personal productivity and wellbeing. This was driven by a range of realities. Firstly, by most measures, I was reasonably successful at work and at home, but I was still frustrated with my productivity. With all this input and the recognition of its importance, I found myself getting more done and more importantly helping others get more done. This was not always labelled explicitly as a simplicity initiative, but those involved will recognise some of the thinking (and passion) that I developed and advocated.

This resulted in *Really Simple*.

In writing this book I was able to bring clarity to my own thinking. The simple act of putting pen to paper (or fingers to keyboard) was to crystallise the various strands of my ideas and establish practices into a more coherent and succinct whole.

So, this is a digest of all that learning and experience, packaged up in such a way that it should be applicable to anyone with a busy and somewhat complicated life, or anyone who feels that they or their organisation would benefit from a simplicity makeover.

How This Book Is Organised

Really Simple is a combination of explanation, approach, and tools.

The explanation details **why** *Really Simple* is important, the approach tackles **what** it comprises, and the tools provide an explanation of **how** to jumpstart your Really Simple initiatives.

- The 'why' chapters (1-3) cover the core thinking, inherent challenges, and typical benefits.

- The 'what' chapters (4-6) cover the model, mindset and method that make up the Really Simple approach.

- The 'how' is also covered in chapters 4-6 through three 'enabling' tools. Six 'support' tools, are also provided, subdivided into 'personal', 'initiative' and 'enterprise'.

Why?	Chapter1: Thinking - *It's Complicated*	Chapter 2: Challenges - *It's Counterintuitive*	Chapter 3: Benefits - *It's Transformational*
What? & How? Tools - *Enablers*	Chapter 4: Model – *It's Multidimensional* RED-CAR: Resources Elements Dependencies Comprehension Appealing Reliability	Chapter 5: Mindset – *It's Psychological* 5i's: Intention Intensity Iterations Intelligence Investment	Chapter 6: Method – *It's Systematic* DASIA: Definition Analysis Synthesis Implementation Affirmation
Tools - *Support*	Personal 1. DF-Plan 2. S-OPRA	Initiative 3. I-Charter 4. DAN-RICO?	Enterprise 5. A-List 6. EA-List

This book provides the fundamental thinking behind Really Simple. More detail and practical downloadable templates can be found on the Really Simple website:

www.really-simple.com

Conventions

There are several conventions in use in the book:

Quotes	**"For a quote" A. N. Other.**
Examples	RSX - For Examples. Some further detail is provided.
Comments	These are points of clarification or specifics that did not necessarily need to be included but the author felt provided helpful and practical context

Really Simple

Throughout the book I use the term 'Really Simple' in different ways:

- as the name and content of the book;
- as the name of the approach detailed in the book;
- describing the execution of the approach;
- describing the outcomes of the approach.

I also abbreviate Really Simple to 'RS' where I think it helps the flow.

Clifford Burroughs

Introduction: It's Really Simple

The power of simplicity has always been recognised by scholars, scientists, leaders, and businesspeople. It helps solve difficult problems, underpins scientific theory and is the basis for some of the most successful enterprises of all time. For others, it just provides daily routine and spiritual guidance. However, whilst some seem to intuitively understand how to leverage simplicity, they also sometimes struggle to articulate how they did so.

Really Simple tackles this by improving people's understanding of the nature of simplicity and providing models, a mindset, and a methodology that can help in both your work and home life. It also highlights ways to counteract creeping complication, which, if left unchecked, can be debilitating to you or your organisation. It applies to processes, services, products, and day-to-day tasks, with the objective of better outcomes and experiences.

Why is *Really Simple* Really Important?

The Really Simple approach is a systematic way of conquering complication. It was developed to enable anyone to create a genuinely simple solution. This can be either where complication already exists or where simplicity needs to be actively 'designed-in'. The fundamental premise is that without a systematic approach you are likely to achieve only a marginal or superficial improvement. Even worse is the possibility that the overall solution might be degraded, even if one part of the outcome feels or seems simpler.

A way to think about this is on a spectrum. This provides a 'relativity' check as to whether you have indeed created something Really Simple.

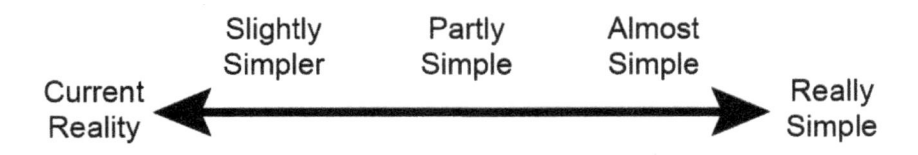

```
                    Really Simple
              Slightly    Partly    Almost
              Simpler     Simple    Simple
   Current  ←─────────────────────────────→  Really
   Reality                                    Simple
```

The objective of the book, therefore, is to give you the right approach to make sure that you achieve Really Simple solutions.

Whilst there are other books on simplicity that describe either principles or successful outcomes, this one provides a systematic approach to creating simplicity at work and in life more generally.

Therefore, this book is for everyone, on the basis that everyone has some complication in their lives. More specifically, this book will be a valuable resource for busy people, managing situations where they feel that complication is holding them back.

The first half of the book focuses on why simplicity is so valuable. It starts by deeply understanding the challenges of complication and why it is seemingly so counterintuitively hard to address and concludes with the potential of the Really Simple approach, and why it is so worthwhile.

The second half of the book is focused on what is required to make Really Simple a reality and how. It provides a specific model (RED-CAR) to help decompose the challenges, suggests the right mindset (5i's), and outlines a methodology (DASIA) to manage the changing lifecycle of specific initiatives.

There is then a specific section on tools that support the development of Really Simple (RS) outcomes. These tools are themselves underpinned by simplicity and have all been developed to solve common complication problems.

Whilst the original catalyst was the inefficient way in which big systems and process projects were delivered, it expanded to address the daily challenges of middle and senior management. This was because, although those I observed were broadly successful, oftentimes this was through experience, guile, and 60-hour work weeks, instead of an inherently productive approach.

The thinking has been refined by observation, research, and experimentation. This was mainly in the workplace but sometimes at home and with a particular focus on the use of technology, both at the individual and enterprise level. The book provides a unique balance of insight and practicality to help you solve your complication challenges in the modern world.

"Do I contradict myself?

Very well, then I contradict myself,

I am large, I contain multitudes"

–

Walt Whitman

One challenge with any book on simplicity is keeping it simple, but also not overly simplistic. The first half of the book tackles the more conceptual understanding, and the second half covers the practical aspects, particularly the tools, which can be easily referenced.

This is not a book about generating more profit or personal wealth, although that may be a desirable by-product. This is a book that seeks to play a part in making the world a slightly better place in the face of persistent complication.

1

Thinking: It's Complicated

The insidious impact on our work and home lives

---

*"Life is too complicated,
so don't take it too 'easy'"*

—

Imtiaz Khan

The world is a complicated place, and this is in many ways a good thing. The diversity and richness of the planet has provided many opportunities for humans, which have been so successful that the population will reach 10 billion by 2050.

> Most people would not view a worldwide population of 10 billion as a true sign of success, particularly related to the environmental impact. I agree there is a problem, but also accept that this is the current reality and Really Simple can help find solutions.

Complication can, however, cause significant problems. Your day-to-day activities in your home and work life all have layers of complication that drive frustration, inefficiency, and stagnation. Being able to find solutions to this is at the heart of Really Simple.

When asked to define simplicity, most people will include the word 'easy' somewhere in their response. They will also often use examples of simplicity centred around minimalism. Conversely, when asked what the opposite of simplicity is, most people will react with 'complexity' and in defining complexity they may use words like 'frustrating', 'difficult' and 'lots of components'.

Complexity and Complication

Complexity is not the same as complication.

Complexity can be defined in several ways, but broadly can be thought of as a significant number of elements that work together to create an outcome that is difficult to predict. There are complex problems all around us, in society, business and nature. Medics, meteorologists, and governments all wrestle with these, embracing the complexity, but also recognising the inherent challenges that exist.

Depending on the context, complexity can be seen as either positive or negative, or sometimes both. Complexity can be very interesting and managing it within parameters can be very rewarding. Google and Facebook's founders did not know (and still do not completely know) the impact they would have on society. Both organisations were motivated to bring some structure to the internet through search and social interactions. Neither could have foreseen implications for politics, society and commerce which are now exceedingly difficult to control.

The unpredictable nature of complexity means that we have a higher acceptance of the frustrations that it might cause. We do not tend to blame the weather forecasters when we leave home without our umbrella and then get soaked in a downpour.

People do however use the term complexity in the negative sense, when in fact they are referring to complication, which is predictable and should be manageable. Complication usually surfaces as a set of interdependent rules, instructions, or steps that those involved consider at least potentially unnecessary, inefficient, or frustrating. Mostly these comprise everyday tasks at work and in life that most of us feel we should be able to tackle without too much difficulty.

Quite often we accept that the consequence of complication may be the need for an expert to help us. A good example is an expert mechanic repairing a sophisticated, modern car. They will have access to training, equipment, and manuals that mean they can fix almost anything.

Perhaps most importantly complexity tends to be inherent, whereas complication is induced, meaning complication tends to be a choice. This book focuses on understanding how complication can occur, how it might be addressed, and how it is critical to recognise complexity, which may overlap or interact with complication.

The schematic diagram below illustrates that variable or regular inputs do, in part, define whether you are dealing with complexity

Really Simple

or a complication. The response is either a 'dynamic' or 'defined' approach with different types of output predictability.

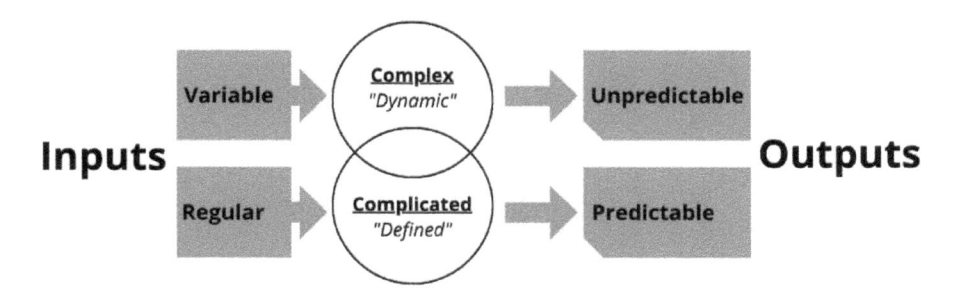

Inputs → Variable → Complex "Dynamic" → Unpredictable → **Outputs**
Inputs → Regular → Complicated "Defined" → Predictable → **Outputs**

The overlapping circles illustrate where complexity and complication can coexist. For example, a modern-day call centre may have simplified many of the interactions, and if not, an expert can systematically steer you through a specific complication. Although, this will still not cover every scenario where a particular combination of circumstances will add an unexpected complexity requiring a dynamic solution.

As I said before, not all complication is bad. The phrase 'good and bad complication' is sometimes used, which can help identify complications worth keeping. An obvious example of a 'good complication' is coffee shops, which offer an amazing range of options with relatively few inputs, and relatively small operational impact.

> That said, as coffee shops have expanded their offerings to include iced drinks and food, the queues have lengthened at busy times, meaning they may be at the tipping point between good and bad complication.

Whether something is considered complicated is to some extent based on individual perception and experience. Usually there will be some common agreement that there is a complication challenge but not necessarily a clear agreement on how to tackle it. Really Simple thinking is focused on helping prevent or reduce complication, which in turn will reduce those negative perceptions and frustrations.

It's In Your Head

Simplicity is an experience. Whether it is a product, service or process, the essence of that experience will leave you with an impression or feeling. Whilst I say this is 'in-your-head' it is the reaction of all the senses that are part of the experience, which means it can even be guttural.

Frustration Claxon

Whenever we are asked to use any quantity of mental energy, our brains do some quick prioritisation. This may start with the basic fight or flight responses, but usually conscious thought takes over to tackle the problem. This means when we experience complication there is an initial reaction which cannot be stopped. This can happen in all sorts of situations, from the most mundane to the very difficult, and changes over time. For example, I grew up in a world where cash was the simple way to pay for something relatively small. Now I become subliminally frustrated at the (small) complication of 'real' money (getting the right amount, managing the change, etc). Conversely, when I am paying with contactless, it has a neutral impact, I just do it. However, if a vendor does not take contactless, I make a small mental note and shop elsewhere.

Great Expectations

The underlying problem that triggers frustration is a pre-wired expectation. If we expect something to be simple, we will not always notice when it is. However, if we expect something to be simple and it isn't, this can lead to frustration.

At some point in your life, your brain will have created some level of prior belief and association that can cause frustration when it does not play out as expected. There are two types of frustration, the prior belief that something *is* simple, and the opinion that something *should* be simple. If you had a simplicity-based belief, any level of contradictory complication will trigger some level of frustration. You might not show it outwardly, and you might minimise it internally, but it still occurs.

For example, the iPhone has changed expectations, particularly if you are an Apple ecosystem user. You expect everything to interact in a way that achieves your desired outcome. Both the Blackberry and the Windows phone had strong capabilities but are now almost dead because they did not provide a complete enough solution early enough in their life cycle. Android is now the mainstream alternative as its developer (Google) focused on providing ease of use, integrated solutions, and app development.

> As of 2020, there are circa 2.5m apps available in the Google Play Store and 2.2m in the Apple App Store. Annually there are 250bn app downloads. This is a good example of simplicity driving success. The ability to search, find and download an app is so easy that the problem of app 'clutter' is probably more serious. Even Microsoft is

> embracing both mobile platforms, recognising that they need to be (a big) part of these huge app ecosystems.

We are conditioned to expect some complication (like completing a tax return), so we get prepared or expert help. What is worse however, is unexpected complication.

RSX (Really Simple Example) - What's this?

Occasionally, you get a letter or an email out of the blue. Very recently my wife received three letters from Transport for London on the same day, with fines for unpaid congestion and emission charges. Coincidentally, I had to pay a fine a few weeks before, so her initial reaction was "What have you done now?" After some significant checking, we realised it was not us, and our number plate had been cloned. This then forced many follow-up actions with Transport for London, the Metropolitan Police, and the Driver Vehicle Licencing Authority. Each call was complicated, involving telephone menu systems and further follow-up actions, emails, and letters, all triggered by something out of our control.

When something like this happens, you really want to press a big button that says, 'My number plate has been cloned' and all the relevant systems and authorities are informed. Instead, you know that you will have to convince everyone that it was not you in a world where you should be innocent until proven guilty. Even then it will often require some follow-up. In this case, it is likely that this must happen regularly, however, it felt like the first time that this had happened to any of the people we talked to. The various humans, web portals, and documents did not seem to even recognise it was a possibility.

Decision Fatigue

One of the key benefits of the modern world is the number of opportunities that life can provide. Making the most of this will be quite different depending on context, but most of us will have daily choices we need to make about what we eat, what we wear, the route we take, the type of exercise we do, the time we spend on anything (including watching TV, browsing the internet, or reading a book.)

If these choices are not simple it will usually trigger some decision making. Do I continue? Do I stop? Do I change my path? Do I completely reassess the situation? You often will not be conscious of this internal processing because it is so ingrained, but it does have a toll. The likely result is that you will make a poor decision, no decision, or miss out on something else. Once you appreciate the impact of complications that result from the decisions everyone is making, you can develop more productive solutions.

The way that most people deal with this level of daily decision making is through routine and habit. In effect, we automate these daily chores. However, it has the benefit of minimising decision making and marginalising necessary but mundane activities.

The bewildering level of product/service choices we face in our daily life doesn't help. This relatively recent (the last 50 years) phenomenon is driven partly by marketing professionals realising that, in many cases, increased variety would give an impression of choice but also, more importantly, different value enhancement opportunities. Think of the long extras list if you are ever buying a new car (on top of the huge range of actual cars and manufacturers), compared to the more day-to-day products (and suggested alternatives) on Amazon, Apps in the Play Store, or selecting a washing powder on a supermarket shelf.

In his book *The Paradox of Choice* (Schwartz 2004), Barry Schwartz explains the paralysis this can cause when faced with an overwhelming choice at the supermarket (he uses 275 salad dressings as an example). More insidious is that even when you have diligently picked something, your satisfaction is reduced by the worrying thought that you have not picked the perfect option. He attributes this to four things:

- your regret that an alternative would have been better;

- the potential opportunity cost provided by the alternative not chosen;

- the reality versus self-defined expectations;

- the underlying reality that in a world of extreme choice there is no one to blame but yourself if you pick the wrong thing!

Maybe more critically it could be a university degree course your child might be considering. This is a huge financial burden for most families so getting it right is really important, and the choice is complicated by the variety of available options.

Barry Schwartz has come to believe that although some consumer choice is a good thing, overwhelming choice is actually a bad thing and is probably a significant contributor to the increasing levels of mental health issues in the industrialised nations. This is so counterintuitive that some people will be shaking their heads as they read this. All I would say is that this is a real problem for a significant enough section of the population that it is worthy of consideration if you make product range decisions. Bear in mind, there will be a hidden cost of providing this choice, be it a personal (usually measured in time) or an organisational impact on manufacturing capacity, pack sizes, stock and, ultimately, profit.

It is certainly the case that these choices and opportunities are not open to everyone and are more prevalent in the developed western world. In the UK, for example, we have an incredible array of utility tariffs to suit every need, whereas these options will not be generally available in all parts of the world. In the UK, it is so complicated that companies exist just to help us sift through the options, helping us decide what tariff is right for us and then support us when switching.

Another dimension to decision complication is what Roy Baumeister and John Tierney highlight in their book *Willpower: Rediscovering the Greatest Human Strength* (Baumeister and Tierney 2012). They describe a limited reservoir of energy that is dedicated to decision making and willpower activity, which is slowly exhausted every day until making the most basic decision is seemingly beyond us. To put this in everyday terms, after a long day, many of us might make less sensible choices in the evening. This partly explains why we graze on snacks and might succumb to a glass of wine or a beer, even when we might be trying to cut down. The key point is that for something to be RS there cannot be too much decision making involved in the product or process.

RSX - Decisions, Decisions

My wife Amanda and I had a long-held ambition to build our own house. Eventually, (despite plots in Southeast England being very scarce) we found somewhere to build. We then had to design within the constraints of the plot and the municipal planning department. This was quite tricky because a) architects that actually follow a brief are pretty hard to find and b) the planning rules were tighter than we expected. In hindsight this turned out to be the easy part of the project. It soon transpired that the number of options of building materials/techniques that were within our sensible build budget were mind boggling: foundation systems, building blocks, cladding bricks, roofing tiles, window types, and garage doors all had endless parameters driving potentially different decisions.

Our builder, Larry, found this amusing but was always respectful of our decisions because he understood we were trying to make smart calls. That being said, he would have picked everything on the basis of something or someone he already knew, reducing his decision complication and the risk of the unknown. All this choice did however mean that we could pretty much get what we wanted (well, for the most part!).

An example of one of the most frustrating problems was when Amanda toiled over a decision about which shower tray to use. We would then run this past the plumber and builder for issues, who would sometimes point out a serious miss on our part (e.g., a plug hole is over a steel joist). Assuming all these hurdles were jumped, we then would reach out to the supply chain, only to find the model we had picked was unavailable for some reason or another.

At this point Amanda was not quite back to square one but it often felt like it. She would go back to the catalogues,

> builder, plumber and call me saying "Here is the alternative, is this OK?". By the end of the build, decisions like light switches and wall colour were made more quickly and with less care as decision fatigue had set in and we just wanted it over. After all, there are only so many shades of white you can compare without going mad.
>
> Obviously, you learn a great deal doing this the first time and we would do almost everything very differently now (although we would definitely use Larry). My guess is we would still make some errors on our second build and would only nail it on the third. The main difference would be our determination to bottom out many more of the seemingly small decisions before we started and look to mitigate the supply chain risk. This would leave us with more available decisional energy for the big stuff.

Some people have come to recognise this is a problem and make quick decisions, without worrying too much about the consequences, feeling that making the decision is more important than its absolute correctness. The wealthier you are, the more likely this strategy will be to work. You just buy a fully loaded car and if you get it wrong, you buy another one. However, for most people they need to get this right the first time, so they will spend time (and decisional energy) trying to reach a fairly good solution.

Different circumstances will cause quite different types of decision complications. Through their foundation Bill and Melinda Gates have what most people might think is the enviable problem of how to invest their great wealth in philanthropic causes. To help with this, they have created some rules, but nevertheless I wager they have moments when they must sweat the decisions because of the opportunity cost of making a poor call.

Action Overload

Many of us live with multiple action lists associated with the various aspects of our lives. Many of our actions are part of a sequence of interdependent actions that all need to be completed to achieve something specifically useful. Some people consider themselves well organised with diaries and to-do lists. However, in my work life I have only met a few people that had a truly robust and reasonably efficient method for doing this.

In the book *Getting Things Done* (Allen 2015), David Allen describes the concept of 'open loops' (pg 14) whereby the brain frantically tries to keep track of all the things that are literally on its mind. His basic premise is that the reflective (thinking) part of our brains are extremely poor at keeping track of everything we want or need to do. Hence you need 'a system you trust' (pg 17) to capture your thoughts and clear your mind, so it can focus on real value-adding work.

Many of our actions are connected to others and Allen labels these as 'projects'. Most of us see a list of related actions without realising they are intrinsically connected. The core problem is that most of us have 'loose' arrangements in place to manage the incoming stream of requests for some type of action, or indeed our regular activity. This is often on top of those things we are obliged to do and any commitments we may have made, either with work or family/socially.

This is part of the mental complication we all face because just about anything you are trying to get done needs to be articulated in some way, then scheduled (possibly with others) and then done, before finally registering as complete. Keeping track of all this and recalibrating continually is mentally exhausting. For those with a strong sense of ownership or standards, you have the additional problem of having too much to do and the anxiety that flows from this.

Trigger Unhappy

Whilst complication is something we all experience, addressing it means recognising the underlying drivers. Within these drivers are 'triggers', whereby you reach an unhappy point where you tip from simplicity into complication. Tackling complication means stopping or minimising these common drivers from reaching the trigger point for those involved. Significantly, these triggers will fire at different times and in different ways, depending on familiarity or expertise. By paying attention to triggers, you can try to reduce the complication threshold across the different stakeholders.

Complication has four core triggers that we need to recognise:

1. Volume

Volume has the most obvious triggers (and the most obvious solutions), where having too many elements can be overwhelming or create an overload. This is relatively easy to see, like an overbrimming email inbox, a process line with many steps, or a long to-do list. However, this is not a problem in all circumstances. A library, for example, needs a high volume of books to meet its core objective. The volume does not cause problems because there is a trusted system of sorting and categorising the books sensibly.

2. Performance

What is slightly less obvious is the ability of something to keep performing as expected. This is because there will often be a lag until this trigger fires, as deterioration occurs. Sometimes there are obvious signs of potential failure, but often these might be buried behind an outer veneer of solidity.

3. Appeal

This will be how something looks and feels, and whether it has some valuable benefit or if it is just useful. These will colour someone's judgement immediately, triggering their perception of complication. This is why estate agents are so concerned with both kerb and threshold appeal. They know that an attractive house will be easy to sell, and the opposite will also apply.

4. Understanding

Being able to do something as intended, and relatively quickly, is important. The opposite is extremely frustrating. OK, it is true that a small subset of the population love tackling something complicated and working out what to do next, but most just want this initial step to be as intuitive and as quick as possible.

It's Multi-Level

Something might be simple at one level and complicated at another.

The core microprocessor in a PC is nowadays incredibly powerful but conceptually relatively simple. The challenge of manufacturing that silicon chip with billions of transistors, which behave reliably, is much more complicated.

You then insert that silicon chip into a computer with a stack of software and other components, each piece of which might be simple but in aggregate is complicated. There is also the potential for complexity as the known limits of physics come into play. The end user now sees a computer as a piece of consumer electronics which is just expected to work, without the need to consider what is happening behind the scenes.

> This is an example of multi-level complexity, complication, and simplicity. The aggregation of billions of simple circuits and software instructions creates the potential for the computer driven capabilities that we have today. As an occasional sailor, the confidence that I get from the now affordable GPS systems, which are both hyper-accurate and reliable, transforms my level of enjoyment; knowing that I know where I am is very comforting!

Time: The Eternal Enemy

It can be a difficult to understand bus timetables, a DIY shop's endless choice of paint, a long visa application form, fraudulent use of your bank card requiring lots of follow up calls, or the gauntlet of your daily commute. Most of this is not in your control and will often be a big source of complication frustration because it consumes unnecessary time.

There are other things that are more in your control but still feel suboptimal. Usually these are in your personal life. These might be the simple things that are nagging away at you, like a fault on your car that needs a trip to the garage, but you keep putting it off. In this type of situation, the frustration is often more driven by annoyance with oneself for putting off something which could have been quite quick to resolve.

Then there are those sub-optimal things, with which you are connected, that you may be able to change. This might be at work or in society. At work, you may be involved in a process which happens only occasionally but remains frustrating. And in society, you may be at a set of local traffic lights that have the wrong prioritisation sequence. You are not powerless to get this changed, but it is going to take time whilst you prompt the relevant civic authority.

In all cases, the same basic frustrations exist, and they usually relate to an impact on how you will spend or have spent your time and the perceived benefit. Our measurement of time and what we manage to achieve over a certain time frame is often how we are judged, but more importantly how we judge ourselves. Time (and space) remain a fundamental human constraint.

As we will see, the Really Simple approach is focused on the root causes of these time-stealers by making as much as possible, as simple as possible.

It's About People

One of the complications that Really Simple recognises is the different roles that individuals play in simplification. This may be executives who conclude simplicity is a key enabler of success and want everybody in the organisation to play a part. It may be for a manager trying to tackle a specific problem, or a proactive individual trying to make a relatively small change to their own day-to-day living. For others, it may be the answer as to why they or their organisation are not achieving their hopes or ambitions. For many (usually customers), something will feel much harder than it should. These varied stakeholders bring different experience, motivation, and skills to any RS initiative, so it is critical to actively create the right blend and accountabilities.

The Four Really Simple Roles

Once you are clear on the objective of your RS initiative, there are four fundamental stakeholder roles required to develop an RS solution.

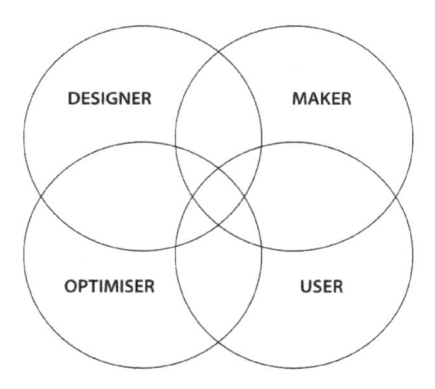

Understanding the roles people play is crucial to ensuring that RS initiatives have the right level and type of resources.

It is worth noting that sometimes people play more than one role in the development of an RS solution. These overlaps are normal but need to be consciously managed to prevent bias away from the objective.

We can look at each of these RS roles in a little more detail. This helps to be sure that, as you embark on an RS initiative, you are clear about who is doing what and that you have the right mix and separation of responsibilities.

1. The Designer

> *"Design is not just what it looks like and feels like. Design is how it works".*
>
> —
>
> Steve Jobs

In this context, a designer is anyone who makes choices about how a thing will exist. This can be a physical thing (often a product), an interior, some artwork, or a building. It can also be a process, procedure, or policy. A designer is a formal job in some organisations and can cover a wide variety of capabilities and connotations, usually focused on some type of physical product.

Really Simple

The reality, however, is that many of us are designing continuously without the formal label of "designer". We make decisions about how we organise something, how we run a process (get something done), or how we produce an output (e.g., a strategic plan or an educational course). Design is therefore usually bundled into one of the tasks we typically describe as management.

Design is a critical activity as it often creates the moment-in-time to lock-in RS thinking. Sometimes this will be the architectural foundations but other times it will be the look, feel, and workings of an object, machine, or process.

In a manufacturing situation, design can lock-in some specifics and will usually be capital intensive. (However, the best designs allow flexibility for undetermined requirements.)

Conversely, the 'soft' design of a spreadsheet, PowerPoint or workflow can be an opportunity to test some RS thinking. This does have a downside, however, when you quickly solve a problem with an 'OK' solution which then limps along and never becomes RS.

A process can also be something personal like processing your email inbox(es), or organising filing and calendars, or something as challenging as managing an innovation pipeline. An RS solution for these can have a significant positive effect on your productivity. It can also reduce the mental drag that occurs with either a complex or bad solution, full of rework, waste, and misunderstandings.

2. The Maker (Server/Deliverer)

The maker role is also crucial but in a different way. A maker is usually the person taking inputs and converting them into outputs. In a factory it will be those who run and support the production lines. In a restaurant it will be the kitchen and waiting staff.

Where a designer can usually create a design up to a point on paper, computer or possibly some form of model or prototype, it is often the deep understanding of the maker that can elevate those designs to become Really Simple. A software engineer, a machine tool engineer, a builder, hotel manager, or schoolteacher usually inherit an initial design and then need to create a working solution. The likelihood of creating an RS outcome will be greatly enhanced by timely and appropriate levels of the maker's involvement in the initial design.

> ### RSX - Designer/Maker Tensions
>
> In the UK there are many TV home and garden makeover programmes. I like these programmes. There is usually a core back story of someone who could use some help, as their home and more general circumstances could be improved by the makeover. The major tension is time. "Can the makeover be done in time?"
>
> The other major (slightly comical and exaggerated) tension is between the designer and the builders. The builders are practical, and the designer is decidedly not. There is plenty of on-screen tussling between the unrealistic designer and the folk that are doing the hard (physical and often highly skilled) work. Occasionally the builder's win, but not often because there is the conditioned master and servant relationship. In the TV shows, this is all part of the theatre but

> in real life this can lead to constant stresses and strains. Needless to say, designers (and other creatives) are often typecast with certain cliched attributes, but probably not more or less than builders!

3. The Optimiser

There are some situations where there is little room for manoeuvre. The design has been done, the big investments locked in, and the status quo established. You might be forgiven for thinking that there is little to be done. However, opportunities in optimiser situations will far outnumber those in design, and more importantly, many of the designs will be far from optimal anyway.

Being an optimiser is a special skill and requires the ability to see beyond how things are done today. It requires finding an incremental simplification that may make a big difference in specific circumstances.

My guess is that a McDonald's restaurant manager (a role of significant responsibility) does not have a lot of scope to design. The menus, the kit, and the physical restaurants will often already be in place. However, the specifics of the restaurant, the attitude and training of the staff, and the clientele will mean there is significant opportunity to tune the operation for maximum performance and safety.

4. The User (or Consumer)

The end-user of anything is probably the most important stakeholder role, as they will usually be the one who judges the success of the solution. In small scale applications, like the organisation of a kitchen or workshop, there will be relatively few people making this judgement. They may not even notice any issues because they are just so familiar with the status quo. Whilst having end-to-end control seems positive, being the designer, optimiser, and user blunts your appreciation of improvement opportunities.

The user will be particularly important in simplification challenges that have mass consumption, such as consumer product manufacture or an App. This can be tricky as they will have requirements both known and unknown, common and disparate.

Known requirements are those that are obvious to the user. For a train booking app this might include a search of the available trains and prices on a particular day. This would also be an example of a common requirement, meaning everyone will need this functionality. A disparate requirement applicable to a minority of users might be making a regular but not daily journey, so needing the ability to 'bulk-buy' tickets with the desire of getting better value. A common but unknown requirement might be a feature to create electronic receipts for expense reimbursements.

Understanding user needs is critical, as getting it right can produce balanced solutions. Also appreciating that users do not usually have a deep understanding of what is necessary or possible means they need some help. (In the case of unknown requirements, it is really the responsibility of the designer to expose likely unknown requirements to prevent big misses.)

Really Simple

Cars are an obvious example. Most users just want to start the ignition and get somewhere, not to understand how the car works. Users will therefore often express their requirements using some form of reference point. An example of this could be someone who knows they want a car that produces zero emissions. They probably know that means electric, which means a battery, but they still want a traditional looking car. This is how you get the Tesla 3. They probably do not say, 'I want a single large touch screen that shows and controls all of the car's sub-systems'.

"If I had asked them what they had wanted, they would have said a faster horse"

–

Henry Ford

Steve Jobs also referred to non-incremental leaps, where a small (usually very skilled and experienced) group of people would have to predict what a wider marketplace would need and value. In the case of the iPod and the Model T, both had a market. What neither Ford nor Jobs could predict was the scale of their success, which became acts of faith given the investment risk they were taking.

RSX - Only One Way to Make a Product, Right? Wrong!

In one of the large foods businesses I worked for, I was always amazed by how much variability there was in the end product. Sometimes this was because the product was produced in different factories on different machines. Sometimes it was because of extreme atmospheric conditions (usually heat). Sometimes it was because of the maintenance regime that made it uneconomic to keep everything in tip-top condition, i.e., some small variation was a price worth paying.

More incredibly (at least to me) was that it could depend on the shift. Usually, these factories were 24/7, meaning that about every 8 or 12 hours a new set of operatives would arrive and take over a production line. At this point, often the previous shift had the machines configured with certain settings, and the new shift would change those settings to what they felt produced a better output (waste, quality, or volume). Then when the previous shift arrived, they would turn everything back. This would be despite SOPs (Standard Operating Procedures) being in place.

The truth in these types of situations is that the method is passed from person to person, rather than a rigorous application of the standard (from the SOP). Like a game of Chinese whispers, the nuances of the method were lost in each human handoff. In this example, there are elements of comprehension, decisional and human complication, none of which were intellectually or independently hard to solve. Most importantly they required a mindset shift, on the part of both the operators and managers, enabling them to start training, doing, and measuring the right things.

It's Not a Silver Bullet (or a Value)

Whilst RS thinking can be a critical enabler, individuals and organisations need to do many other things in the right way to be successful. They need to manage their stakeholders, their finances, their quality, their marketing, and their operations. The RS approach can be applied to all these areas, but other types of thinking will be just as critical.

RS is a Way-Of-Working (WOW) akin to deciding that the "Customer is King". Both WOWs really refer to the type of work that you believe will create a positive future and, therefore, should be a strategic theme or principle. Really Simple should, however, not usually be considered a value. In contrast to a WOW, values tend to be words that imply behaviours and are the shared emotional (human) approach, such as innovative, progressive, and energetic. They should imbue the behaviours you expect from individuals and teams and allow management to drive the organisation in the right direction.

> Values need to reflect what is specifically required to make your organisation the best it can be. Sometimes with values, I have seen words like 'honesty' and 'respect'. These are usually tickets to the game of life and are almost contractual. Even the very biggest organisations can get this confused, and some will not worry about the semantics. However, being clear about the differences between values, principles, objectives, themes, visions, missions, and purpose will speed up the process of defining and communicating them in an RS way.

2

Challenges: It's Counterintuitive

It's not easy

---

*"To be truly simple,
you have to go really deep"*

—

Steve Jobs

One of the challenges of the word 'simplicity' is the priming effect that it has. Priming is a psychological mechanic of the brain whereby one word or image creates an association with another, usually a synonym. The result is that simplicity is associated with easiness. This in turn risks the creation of simplistic solutions.

Being simplistic means being *overly* simple, potentially to the point of an extreme, and usually with too low a level of solution specifics.

---

*"If all you have is a hammer,
everything looks like a nail"*

—

Abraham Maslow

Really Simple

A simplistic solution can be tempting as often you look like you are making progress whereas you are really creating other problems that will in turn need to be managed. The solution, often driven by time pressures, might even be so bad that under challenge it proves to be a retrograde step.

The second key problem is that making something simple is harder to do than everybody thinks. Usually, the necessary decisions will require energy and courage.

RSX - Less Is More

One of the most famous simplification stories is that of Steve Jobs arriving back at Apple and culling the product line. He completely stopped marketing a hand-held computer called the Newton, slashed the product lines, and divided the new offerings into four quadrants (see below).

	CONSUMER	PRO
DESKTOP	iMac	Power Mac
NOTEBOOK	iBook	PowerBook

This meant job losses and role changes for some legendary Apple engineers. It meant walking away from some cash

> generating business. This was not easy for Jobs or Apple employees. He was intense, demanding and needed to win some battles, and in the early days it was far from a sure thing that Apple would make it. But Jobs wasn't playing the percentages, combining some hard-won experience (Apple, Next and Pixar) with the belief that Apple's success was based on it being different.

It needs to be said that this quadrant representation was slightly simplistic because there were several sub-options, meaning it was more complicated than it looked at first glance. It is also worth noting that Apple was struggling, and bold action was required. Steve Jobs managed to bring focus to the whole company and a sense of individuality to the products. Most of the technology existed but it needed to be synthesised into the right combinations and with the right branding. Engineering simplification was also critical, enabling the products to be made, shipped, sold and supported in a profitable way.

At this point Apple was like many organisations, stuck somewhere between surviving and doing OK. Like many others, they were unable to see their complications and the application of standard business doctrine was failing. Only a completely external perspective combined with courage was enough to shake them out of mediocrity.

Beware of Nonchalance

We are all familiar with the phrase "It should be simple to…". It is used in daily life to articulate our view that something could be easier. We tend to use this when we are dealing with something that is repetitive or has longevity. This can engender a level of nonchalance or acceptance that is not helpful, when the reality is that there is some underlying complication that needs to be tackled.

It is both amazing and logical that many people don't even realise they have a complication problem. They jog along, doing what they have always done or what they have been told to do. They may complain about some aspects, but mostly they feel they are lacking in the ability to change anything, so they just carry on doing what they have always done.

The Particular Significance of Priming

Our experiences, combined with the brain's inbuilt bias to take shortcuts, means that people will often substitute a common word with another. Knowing this is useful in all walks of life, as it means that it is possible to nudge people in a sensible direction both subliminally and explicitly. In the context of RS, it can be slightly problematic, as it creates an unhelpful nudge.

The word 'simple' can be misleading and have negative connotations. In the case of 'simple', the next thought that can leap to mind is 'Simon' (a nursery rhyme), 'easy' or 'stupid'. Other people that become involved in simplicity initiatives will subconsciously make these associations. However, as a simplicity expert, it is important to realise that you will not have these same reactions and will need to allow time for others to tune in.

Priming is a common psychological tripwire to be aware of. In his bestselling book, *Thinking Fast and Slow* (Kahnerman 2012), Kahneman tackles priming, as well as providing a real-world view of the effect of heuristics and biases on our reactions and behaviours. Understanding priming is specifically important in delivering an RS initiative, helping us spot biases that may lead us in the wrong direction.

The Motivation Problem

In some cases, people just do not have the motivation to simplify things. This can be for several reasons. In the workplace employees are conditioned to assume they are paid for their labour and not to improve what they are doing, and unfortunately some (poor) management practices even reinforce this. In other situations, the culture of an organisation can be driven by top-down thinking, marginalising the importance of bottom-up contribution. While all organisations need some top-down leadership, disenfranchisement of those closest to the complication challenges will create an unnecessary opportunity cost.

In the workplace, this is partly a vestige of the industrial age, and to some degree the business school age, where work specialisation and automation has meant that labour is often the glue that joins together the automation elements.

Despite being important, challenges in the home are often much smaller and largely in our control. Therefore, we will often jog along living with the status quo because our simplification energy might have already been sapped by work-related events not entirely in our control.

Not Making the Payback Connection

One of the motivations for making things Really Simple is to make life easier and to spend more time on the things that make a difference. Seeing a payback, usually in time, when I invest in an RS initiative means I am always looking for opportunities. If you do not make this payback connection, then you are less likely to see the problems and potential solutions.

> ## RSX - A Place for Everything, Everything in Its Place
>
> One easy-to-understand RS example is the layout of a kitchen. This can be a domestic or professional kitchen, where the big items (cooker, sink, fridge, etc.) will be in a specific 'fixed' location. There is then the 'flexible' locations of the utensils and ingredients. These 'flexible' items might be accessed hundreds of times a day, yet there may be an elementary swap around to make this more efficient. In my own house, we have located the dishwasher close to the major items that are washed to make emptying as easy as possible. We also put the regularly used cooking utensils closer to the cooking area. They started the other way around for no other reason than the order they came out of the packing boxes! This small adjustment will mean hundreds of hours saved and more importantly, makes this necessary chore just that little bit easier.

This example minimised motion as moving around the kitchen for the correct utensils was a repetitive task with no value added; it was something that just needed to be done. In a professional kitchen, where space is often limited, this will be a critical enabler of success. It will enable efficient food production, better health and safety, easier training and more, all in the service of a profitable business. The best chefs and restaurant businesses know the importance of this and put significant effort into getting it right.

Change Is Hard

An old boss of mine often quoted the mantra that the strongest species did not thrive as successfully as the most adaptable. Change management, in an organisational context, is one way to stay relevant. This is particularly significant when embarking on simplification. By definition, a RS initiative will involve some form of change. This might impact individuals, processes, and short-term performance. All of these can sow seeds of doubt as to whether the disruption is worth the cost. This will often manifest itself as a staying-power challenge, where the benefits may be clear but sustaining the change will be particularly tough.

The Risk of Complications Gone Bad

There are numerous examples of good complications. The internet, the mapping of the human genome, and air traffic management all provide benefits that reshape how we live. These life changing wonders are often based on a few simple fundamentals that have been elaborated by a team of engineers to create something incredibly useful. The problem only comes when they add too many bells and whistles meaning that complication starts to set in.

Microsoft has long dominated the office software market comprising Word, Excel, and PowerPoint. These tools transformed the way we worked and massively simplified the work that we now take for granted. During the 90's, they had

Really Simple

been engaged in a feature/usability war with WordPerfect, which by the beginning of the next decade they had largely won. In the mid-2000's however, their user interface research team realised that people were often only using 10% of the available features because the options were so buried inside the menus.

This legendary image [below] shows all the toolbars that could be opened in Word at the same time.

The 10% usage insight led to a 'more is better' solution. The engineers and user interface designers made the key features more obvious and accessible developing the now pervasive Ribbon interface (below). This stopped the stacking you see in the image above and added text to the icons in a tabbed interface. The small downside was that the Ribbon used quite a bit of screen space, but they recognised this and so provided an auto hide/reveal solution.

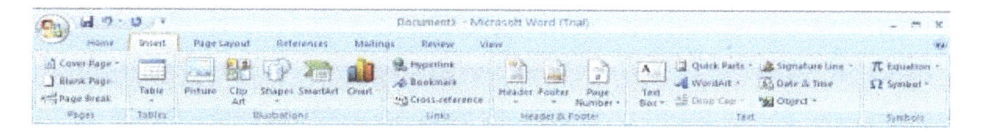

In the very latest incarnation, the design is now moving back towards a more minimal look, that we see here (below), which is a mix of the two previous approaches.

Inherent Difficulty

The Complication Can Be Hidden

If you accept the "it's not easy" challenge, one of the problems is surfacing the complication.

> ### RSX - Understanding What Lies Beneath
>
> An amazing thing about modern computing is that complication is hidden. Apps on smartphones, Applications for PCs, and Websites on Browsers shield us from the underlying complications of the devices, infrastructure, and engineering. The investment of effort and brainpower which enables this veneer of simplicity gives a false sense of how easy it will be to enhance or change. If you use your computer or phone as a consumer, you just want it to work. However, when something goes slightly wrong, the underlying complication makes fixing the problem difficult.

More day-to-day difficulties are the steps necessary to achieve even the most trivial example.

RSX - Getting the Bins Emptied

We are lucky that our local council provides a good range of doorstep recycling collections. We are in a location that has a large amount of garden waste, so we pay for a fortnightly collection service. This service must be renewed annually. However, despite my wife trying every-which-way for this to happen automatically, the system does not process the application. My wife gears herself up for this annual problem but still lives in the hope that one day they might get it right. She tries to forensically help the person on the other end of the phone to address the root cause of an erroneous postcode, but nothing works. (Our best guess is that our address was removed from the system whilst our house was being built and has not been put back correctly. However, trying to speak to someone who can fix this once and for all has escaped both our capability and, to some extent, our patience.) Every year we resort to a hack with the person at the other end of the line, knowing that we will be going through this again in 12 months' time.

There are countless other examples, particularly buried deep in big software systems, of functionality or data that is not understood. Therefore, the hidden complication needs to be consciously sought out. I remain utterly amazed at how often I meet people who are doing something inefficiently without enough perspective to attempt to challenge the status quo.

> ### RSX - Humans Are the 'Glue' That Makes Things Work
>
> The use of software applications is now pervasive, but often they still need to be supported by a help desk, service desk, or response line. You may find yourself phoning a call handling system that will attempt to target you at the right department. Assuming you get through to the right place (which is not always easy), the individual is often providing the 'glue' to make a set of processes, policies, and technologies that are quite complicated feel simpler to the caller.

Concreted In Complication

Whilst entrenched complication is not always an issue, most of us will have witnessed well-established processes and outcomes that are far from ideal. These are often so normalised that even changing to something radically better, that requires only a small amount of energy, will be paralysed by risk aversion – i.e., not disrupting the status quo. This sensitivity to change does vary around the world, but customers are rarely impressed when something that used to work now does not!

However, making changes when the risk is manageable or contained seems insignificant against the challenges of big or long-established organisations. Modern systems and industrialisation have been around for about 125 years. Over that time, organisations that were originally enabled by paper records and postal services have embraced the information age, including Mainframes, Minicomputers, Personal

Computing and then the Internet, the World Wide Web, and most recently Smartphone and Digital solutions.

> ### RSX - Rumours of COBOL's Death Have Been Greatly Exaggerated
>
> COBOL (Common Business-Oriented Language) was created in 1959 as part of a US Department of Defense stopgap to control the escalating costs of software development across its quickly growing computer estate. Its use exploded throughout the '60s and is still the backbone of many of the world's biggest financial systems. As recently as 2017, IBM reported 92 of the top 100 banks still used mainframes for their core businesses. Financial service providers still use COBOL on mainframes because it is fast, efficient, and resilient.
>
> During the '60s, '70s and '80s, COBOL's growth was underpinned by its simplicity with just 43 commands, 87 functions and a very rigid structure. Over the last 40 years, this functionality has expanded, and solutions have been found to allow other software to interface with these, which has extended both its relevance and life.

In huge companies, disruptive changes to their markets mean that this is an almost insurmountable challenge. Kodak, Blockbuster, Nokia, MySpace, and Blackberry have all had the world at their feet and have struggled to reinvent themselves because of the impossibility of moving past their legacy revenue streams.

> ## RSX - IBM Tackles a Structural Strategic Challenge
>
> In 2004, when Sam Palmisano demerged the PC division of IBM, it was the start of a strategic journey away from a business dominated by hardware to software, and then services. At first, it seemed counterintuitive to remove a consumer division of the business that they were well known for. However, Palmisano realised that PCs were becoming commodities and their ability to add further enterprise value without significant investment was severely constrained. He was able to sell the business unit to Lenovo, at a respectable price and take a stake in that partner. As a result, IBM continued to sell ThinkPad laptops to provide the end-to-end service its business clients were used to, without the distraction of competing with Dell, HP, Compaq, or Apple consumer devices. Palmisano simplified IBM without any cost to its fundamental strategy, focusing instead on its core vision of solving its client's toughest challenges and innovating for a smarter planet.

These big organisational realities are often so 'concreted in' their ways that it takes a metaphorical jackhammer to break out of the complication. Therefore, it often becomes necessary for boards to fire CEOs, even when they are reasonably successful. The most important step is recognising this and not underestimating the creativity and energy required to effect the change.

Complication Is Incentivised/Revered

One of the ways in which societies incentivise innovation is with Patents. There has been an inexorable increase in these over the last 50 years with a 73% increase in yearly filings between 2008 and 2018 (Source WIPO 2019). Coming up with something unique and defendable is how some small fortunes have been made.

Obviously, a high level of innovation is not necessarily a problem in and of itself. If innovation is the foundation for your success, it is important to resist the temptation to add complication because it is safer and easier than more fundamental innovation.

Legacy Makes Money

People are often hired and paid well because they know how to cope with the complications in their field of work. Therefore, complication is almost encouraged. This might be a commercially skilled individual that understands the dynamics of a market, or an engineer that has deep experience of a particular piece of equipment. Whilst these individuals or teams might know how to simplify their situations, they have little incentive to do so or share their knowledge.

Some companies even exist to serve this market, taking your complexity away from you. These companies are commonly referred to as outsourcers or out-taskers. They specialise in coping with complication, or indeed adding layers that give the impression of simplification. Usually, they are particularly good at ensuring that the human component is very consistent and productive, with strong measurement and management of individual performance.

Getting Past Subjectivity

To some extent, simplicity is in the eye of the beholder. If you raise the bar, then there is a risk that getting agreement is even harder. This can be a challenge at all sorts of levels. In the home, different family members might have different approaches. In the workplace, there may be many different stakeholders with different perspectives on what would be Really Simple. These subjective opinions usually require a 'tie-breaker'; which is inevitably a senior manager who will need to evaluate the various inputs. Ken Segall brings this to life in his book *Think Simple* (Segall 2016), in the chapter 'Simplicity Loves a Leader', as he recognises that some leaders are spectacularly successful at driving for underlying simplicity.

Most organisations have their problems, sometimes significant, sometimes temporary, sometimes structural. Often the CEO will be rotated fairly regularly by boards and shareholders in search of an improvement. Usually, there will be a bit of a positive bump in performance as the organisation addresses some of the basic problems, perhaps with key people, product quality, or cost control.

Things can stall as the entrenched complications that have not been addressed can start to emerge. The new CEO at this point has a difficult decision because they know that facing up to this will absorb significant time and personal credibility risk. Therefore, they may pull even harder on the classic levers that have helped them thus far. Unfortunately, these levers may bow under the strain of having been already pulled, often by a predecessor. It is easy to lose count of the number of newly appointed CEOs arriving at an organisation and their first activity is 'to take out cost'. Costs (usually meaning people) need to be controlled, of course, but structurally. The success of an organisation will always be better served by objectively removing fundamental waste through efficiency and simplicity.

Really Simple

Complication Stacks (Jenga)

In software, there is often a hidden "stack" of code that sits behind many of today's software-enabled solutions. Developers may be reluctant (or recognise the impracticality) of metaphorically opening up the bonnet and fixing some underlying limitation. Quite a lot of this code will have been written long ago, potentially provided by a third party or may have been developed and made freely available (commonly known as 'open source' code).

Recent software failures at banks and airlines are almost certainly partly due to this, where the level of complication is literally incomprehensible. Those who do have some understanding are, at best, playing a risk management game. But what is probably more worrying are the leaders who do not understand it themselves, wanting to believe that the techies will have it all under control and, if they don't, at least they can blame it on 'a software problem'. This defence does not really work in the modern world, claiming technology ignorance as a CEO now has the equivalence of saying you don't have any people skills. I know some managers will be rolling their eyes at this statement, saying what can be more important than people skills and I agree with them. My point is that managers would never be promoted if they did not show core competence in people skills and that should also be the same for technology.

Clifford Burroughs

Natural Instinct

The Human Condition Is Additive

Almost everyone has been conditioned to believe more is better. In the modern world, we are bombarded from an early age with consumerism, mainly through the various media channels, including journalists, advertising and more recently through vloggers and other social media. More experiences, more stuff, and more money is fashionably encouraged.

We may not wish to respond to these influences, but we also feel the need to stay current in a competitive world. This may mean that we want, or feel we need, things that increase speed, including air travel, computing power, and next-day deliveries. We are likely to see innovations as 'necessary' improvements, like 'whiter whites', in-car navigation, and High-Definition TV. There is also the tendency to believe bigger will be better, so cars, ships, and buildings have all grown. In the situations where miniaturisation has also been important, it has never seemingly been at the expense of increased performance, the Smartphone being the most obvious example.

Really Simple

Clearly, speed, sophistication and scale are not inherently bad and have enabled some significant lifestyle and business opportunities.

> There are obviously some deep sustainability counterarguments that question whether this technology-driven progress is really progress. All opportunities need to be exploited responsibly and that is one of the key roles of good government (of whatever political hue). Governments need to be getting ahead of the sustainability challenges by insisting that primary inputs are sustainable. Having spent some time looking at this in mass food manufacturing, it is a bit of a minefield, but I am always mindful to never let perfection get in the way of progress.

Even more than our propensity to be influenced or our need to "keep-up", we have a primal and competitive instinct to acquire anything that supports our individual survival. In the modern day, this means acquiring knowledge and skills to cope with the complex world, usually with the underlying aim of getting a job or nurturing a family.

We are also 'gatherers', leading to many of us having any amount of paraphernalia in our private lives that we don't necessarily need, but we keep in case we 'might' in the future. (If you are reading this and you are one of the few who are relentlessly organised, throwing out almost everything you are not using, you may conversely want to consider if you are overdoing it.) In the home, clutter is not a disaster, as many can see past the 'distraction' and need to keep it 'just-in-case'. However, if you are an organisation with layer after layer of retained elements, this can be crippling.

Organisations Are Also Additive

An organisational problem will often be identified by the absence of something. This might be too little planning, control, innovation, engineering, promotion, resource, customers, or products. Whilst this is not a universal law, the usual response is to add something new. This might be a new layer on top of something or (slightly better) a replacement for something that is not working. Either way, the number of elements that the organisation comprises usually goes up, or if they do come down, it will be by less than the change might make possible.

Of course, in some cases, the removal of something specific may become a key objective because an aspect is explicitly not working. This might be an individual that is slowing progress, a machine that has become inefficient, and in the most extreme cases, it might mean the sale or shutdown of a complete organisation. A truth is that most organisations have a finite life unless they steadily reinvent themselves.

Businesses Are Additive

Most businesses exist in a context shaped by a competitive marketplace and demanding shareholders, both usually expecting ongoing and continued growth of one type or another.

This, therefore, means that management targets additional sales and profit usually in the service of shareholder value. There is also a phrase which has entered common parlance in business: "If you're not growing, you're dying." In fact, the phrase was originally coined to emphasise the importance of personal development and is now used across the business ecosystem.

This growth imperative has some obvious societal benefits. The competitive (capitalist) model has produced many of the great solutions we now have to real-world problems. These include medical advances, plane travel, and agricultural solutions. In some cases, however, the profit motive becomes so all-encompassing it drives businesses to do the wrong things and the implications for simplicity can also be negative and insidious. We probably don't need 35 different types of olive oil!

Loss Aversion

It seems to be the case that we are all wired to be more 'loss averse' than we are 'gain positive', meaning we are more likely to be unhappy that we have lost £100 than we are to be excited that we have gained £100. Simplification will often mean removing or changing an element of the solution and it is counterintuitive to remove something that seemingly provides a benefit.

Reports, Reports Everywhere

In corporate life, there are often many (Excel) reports being produced and sent around an organisation to support decision making. These reports often flow into tens of email inboxes and often are not read, or if they are, only a few specific data points are referenced. A simplification or lean initiative will quickly ascertain if this should be a target for reduction. However, there is seemingly always one user who uses the reports, so their production continues, and good intentions are lost to loss aversion (unintended pun).

Over-Engineering / Elaboration

If you are an expert (or the person charged with providing a solution is an expert), then there is always the risk of over-elaboration. This is usually because the solution is something that the individual will interact with continually. Whilst these additions may have some unarguable benefits, this is not the same as providing the optimal outcome for everyone else.

100% Solutions: Designing for the Tail

Connected to over-elaboration, is the challenge of staying focused on a sweet-spot and serving that well.

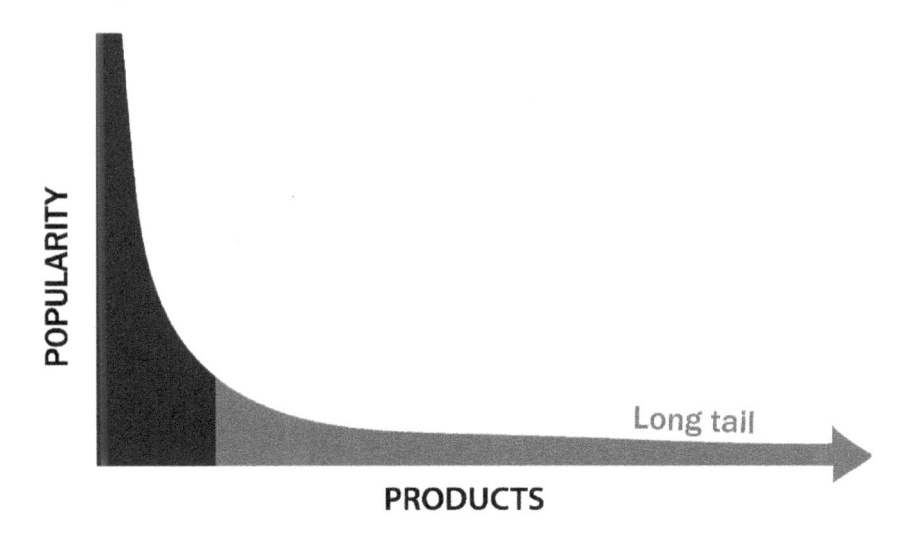

In most organisations, there is a legacy of goods and services, which are often known as the 'long tail'. Connected to the 80/20 (Pareto) rule, many organisations make the bulk of their revenue from 20% of their offerings. The long tail metaphorically describes the physical shape on a graph, showing the revenue vertically, and ranking the product from the most to the least. It is driven mostly by the belief it will add sales but sometimes as part of an attempt to offer a full range or service.

Cutting the tail is often thought to be the holy grail of companies with this type of profile because the standalone profitability is usually terrible, but so is the organisational distraction. Reducing the tail is usually hard, partly because it can reduce the overall sales of an organisation, but often because it strains relationships with customers who are making money from products that you are not.

Being Inspired

As we have seen, designers often take a lead role in simplification initiatives. Typically, people with these roles will have a creative background.

The fundamental hypothesis of this book is that you need to take a systematic approach to simplicity, but this is based on the premise that we need to allow maximum room for creative thinking. However, this is too important to just be left to the designers alone.

It remains true that finding RS breakthroughs will often be through perspiration rather than inspiration. However, for most, the thought that we have had a blinding moment of creativity will feel more special. The key point is that you need both inspiration and systemisation across all roles, and it is often hard to separate the inspiration from the systematic.

3

Benefits: It's Transformational

Untapped and necessary

"The greatest ideas are the simplest."
-
William Golding

We have seen that simplicity needs to address complication and is not as easy to achieve as we might expect. Here we see both the potential and imperative that simplicity can provide by understanding the dramatic impact it can have in work and life.

Brands

Siegal+Gale are a brand agency with a particular focus on simplicity-based experiences. As part of this, they prepare a Simplicity Index to understand the economic value that simplicity creates. They surveyed over 15,000 people in nine countries to understand those providing 'the simplest experiences' on the premise that it is sought after everywhere. They hypothesise that simplicity drives trust and loyalty, and that drives recommendation and financial gain.

Really Simple

In the 2018-19 report, they compared the performance of the publicly traded members of the 'global top 10 simplicity performers'. They calculated that since 2009, these top ten outperformed the big four markets by 679%! (They themselves occupied a reasonable growth range between 52% and 135%). Top of the list was Netflix, followed by Aldi, Google, Lidl, Carrefour, McDonald's, Trivago, Spotify, Uniqlo and Subway. (Obviously to have a chance of making the list it helps if your brand is a success in big markets.) As important as those at the top of the list are those industries that are consistently at the bottom of it, such as insurance, telecommunications, and car rental.

A word of caution here. Obviously, a Simplicity Index is a perception-based survey. Whilst this is good for understanding what is successful, it is less good at highlighting the underlying drivers of success. Trying to understand the ingredients that certain brands mix to make them successful against this index is helpful as a stimulus but will not necessarily map to your situation. Furthermore, although the stats in the Simplicity Index are staggering, within the overall index there are organisations that struggle from time to time. When looking for proxies of success, it is worth remembering that the seminal *In Search of Excellence* by Tom Peters introduced 43 companies that he thought were excellent. However, only a few years later many of them were struggling. Although those companies have still outperformed the market, it has not been at the spectacular levels of those in the Simplicity Index.

> For those of you saying to yourselves that if you have organisations like Google and Netflix included in this index they will warp the statistics, you are somewhat right. The bottom line here is that consumers identified these brands as simple and the hypothesis is that their growth, versus their competitors, was centred on simplicity.

The conclusion is that the modern world values simplicity and will either pay a premium or show loyalty for brands that deliver on this driving good or reliable profitability. For clarity, these brands need to deliver actual and perceived simplicity rather than a veneer.

Communication

Communication is often the silver bullet. For all the brilliant ideas and learning that the modern world provides, if a 'listener' does not understand or engage in a certain situation, the moment will often be lost. Politicians, CEOs, parents, teachers, influencers, to name a few, all come to understand this and use some of the thinking behind RS to underpin the varied types of communication. Therefore, developing your understanding of the nature and value of this is a critical enabler of the RS journey.

First, One-Way

The printed word played a key role in the development of the Renaissance, the Reformation, the Age of Enlightenment, and the scientific revolution. This laid the foundation for the modern knowledge-based economy and, most importantly, spread learning to the masses.

The universal postal services that most countries provide is the ultimate example of simplicity. Write a postcard, add an address and a stamp, post it, and somehow it will arrive quickly at its destination. Behind the scenes there is some degree of complication, but that is hidden from the end-user. Being able to rely on this system has provided a platform that everyone can use either in business or domestically.

Next up came the use of radio waves. This initially allowed telegraph communication using morse code and then broadcasts, with the ability to talk to the nation 'live'. Whilst, at the time, this seemed like scientific magic (words out of thin air), the physics and principles behind radio communications are extremely simple and reliable. It is true that the equipment required to broadcast and receive did have both complication and reliability issues initially. However, the ability to tune into Arthur Burrows reading the first ever BBC news on the 14 November 1922 must have been a revelation (notwithstanding how few people would have had a radio set).

Second, Two-Way

The ability to talk in real time via a telephone was another game changer. It may seem restrictive now, in the sense that someone had to be at the other end to pick up. However, the ability to speak to anyone anywhere in the world transformed politics and commerce. Again, in the beginning, this was relatively crude technology based on the physics of waves over wires, but its simplicity allowed the development of a huge infrastructure of wires, cables, and exchanges that have also stood the test of time. Those cables and wires helped start the internet revolution through dial-up modems and basic forms of broadband over copper wires.

Whilst mail services and landline telephones remain relevant, both have been eclipsed by email and mobile cell phones.

Third, Multi-Way

The simplicity and immediacy of email transformed the way businesspeople, in particular, communicated. However, this is now being usurped by super-simple chat apps, which provide text, voice, and video conversations across groups, in almost real-time. Email now seems quaint in the world of 4/5G, broadband over fibre, video conferencing, and messaging apps.

Global pandemic lockdowns have also ushered in a new wave of video conferencing functionality. Compared to the world of significant daily commutes and business travel, these apps make it possible to increase communication remotely, and break-up large complex topics into smaller more accessible chunks.

Sometimes Emotional

There is a beauty in this quote from JFK.

"Ask not what your country can do for you but what you can do for your country?"

—

JFK

Really Simple

Its simplicity, symmetry, and call to action are so intertwined that it redefined the standard for US Presidential inaugural speeches. Other icons such as Churchill, Gandhi, Mandela, King, Ali, and Dylan have all spoken with simplicity and eloquence and have changed the course of history. I personally remember schoolteachers making profound points with deft precision and there are many memorable quotes that demonstrate the power of simplicity to make an emotional impact.

This also applies to the famous slogans:

- Just Do It
- I'm Loving It
- The Ultimate Driving Machine
- The Real Thing
- Think Different

These can evoke emotional and functional images that underlie our impressions of the brand, but critically do so in very few words.

Often, it's not even verbal. Here are the iconic global logos that convey brand identity:

These logos convey a lot without comprising very much. In fact, it is almost the absence of something that defines these brands; they all have a certain graphical minimalism.

One of my favourite examples of this is the FEDEX logo, which has won many awards for its clever use of the negative space. The hidden arrow in EX is supposed to represent 'speed, accuracy, a strive for perfection, and perseverance in achieving goals'.

Repeatable

It is a simple truth that we are bombarded with communications every day. To have a chance of getting through, they need to be simple and repeatable. Martin Luther King's 'I have a dream' speech worked because it had simplicity but also created a strong memorable theme. The actual speech was quite long and the 'I have a dream' section relatively short, but that repeated and repeatable line is what will always be remembered.

Decisions

It turns out that decision making, in fact all intentional mental effort, is quite testing on both our physiology and our mental capacity. We all intuitively know this. At the end of a testing day, when we make many decisions, we often feel 'wrung-out'. We are desperate to get into bed to 'reboot' ourselves and 'go-again' the next day. Reducing unnecessary decision making is part of RS and creates space for other (more important) things.

> **RSX - Any Suit as Long as It Is Blue or Grey**
>
> It is difficult to deny that any US President is busy. President Obama, like most, was incredibly aware of the power of presentation and knew that picking the right wardrobe was important. However, he consciously decided that he was not going to invest any significant time in it, so every day he would largely get up and put on the same clothes, almost like a uniform. Jeff Goldblum refers to this when playing the scientist in the film 'The Fly, where he had 5 different sets of identical clothes in the wardrobe, arguing that even thinking about what to wear that day was wasted time and energy. Now I know that I am a man writing this book and any female readers may argue that the rule worked for Barack and not Michelle, and I accept that, but the thinking still holds true.

Most importantly, you need to measure this benefit in terms of the opportunity cost. This means measuring the time saved not deciding what to wear is dwarfed by the importance of what you do instead. It is also important to remember that any single decision alone may not cause decision fatigue, but rather the compound effect of making a number of decisions during any one day. Being clear on what is important and reducing decision making of the unimportant is critical.

Products

There are many examples of simple products but relatively few have changed the world as much as these three.

Pencil and Paper

These are so ubiquitous and cheap they almost do not feel like a product. However, despite many attempts to move to digital solutions over the last 20 years, pencil and paper remain the main staple of pupils and businesspeople all over the world.

The reasons for the timeless popularity of pencils and paper are many. This might be creatively, practically, or emotionally; it might be the most important product in history, (assuming you exclude the toilet roll!). Our brains are wired from an early age, through play (colouring in pictures), to handle pencils to create and transfer our thoughts to paper. As we get older and write notes or paragraphs on paper, the inherent simplicity continues not to be matched by electronic substitutes.

The Internet

Most people do not know how the internet works. It is a set of electronic message standards and protocols, which in aggregate are quite complicated (with some elements of network complexity), but independently are relatively simple, building on the basics of telecoms and computers.

Until the beginning of the '90s, wired networks were relatively scarce and specifically designed to link up large organisations. Then the advent of the dial-up modem showed the possibilities of being online to the public. However, it was the advent of ISDN and then modern-day broadband that transformed the reliability and speed. This, in turn, spawned the dot-com era when low-cost airlines, online retailers and internet banks all exploded. These disruptive technologies were mostly made possible because of its architectural simplicity.

This simplicity (and open standards) has resulted in an explosion of internet enabled investment and capabilities. In fact, these architectural foundations are simple enough to cope with the huge quantity and point-to-point messages being transmitted every second.

Barcodes

Like many inventions the barcode was invented many years before it gained widespread use. Originally patented in 1951 in the US by Norman Joseph Woodland and Bernard Silver, the idea was based on morse code with thin and thick printed bars.

After a few false starts, it became clear that barcoding consumer products had significant commercial benefit. Products could be scanned through the supply chain, but particularly at checkouts. It also meant that products could be priced on the shelf rather than individually, reducing labour and increasing pricing flexibility. However, it would only work if enough manufacturers adopted it and those involved used the same standard. These were agreed at the beginning of the '70s, and the big cash register and IT companies started to develop scanning solutions.

Mass adoption occurred through the 80's, first in big supermarkets and then with small shops. Now barcodes are used in distribution and inventory systems all over the world, a testament to the success of this fundamentally simple idea. Today, consumers can even use barcodes to self-scan their shopping and count calories.

Services

Amazon

Amazon is really a service company, underpinned by technology and distribution assets, which now sets the standard for all other retailers. The website and apps are their shopwindow and manage to allow navigation across a mind-boggling range of products. Finding and ordering things is incredibly simple and exceptionally reliable with small crinkles only showing up occasionally. In truth, they are so successful because they have made what could be complicated into a simple service, delivered simply.

First Direct (Branch-Less Banking)

First Direct is a UK based bank which was launched in 1989. At the time it was labelled a 'telephone' bank, although they have added text, web, and app banking over the years. Perhaps it would be more accurately described as a branch-less bank, although you can use HSBC outlets for physical (cash/cheque) transactions. You don't really have a relationship with a bank manager either, which even in 1989 was already unnecessary and slightly outdated. Instead, you phone a number and get personal and competent attention

24/7. I have first-hand experience of the service and it is impressive, consistent, and even friendly.

Underpinning this is a focus on simplicity. The ability for the consumer to ask a question and get something done in a timely and efficient way is critical. Plus, the team is empowered to be able to use discretion. For example, I needed to do something that they do not normally do, but the team found a way of making it possible. This will keep me as a loyal customer and share my positive experience with others.

> First Direct are also simple in more everyday ways. They pioneered the use of laser-printed A4 single colour (black on white paper) for all paper-based communication, which provided flexibility, consistency, and brand congruence.

Search

Sometimes it can seem hard to remember what we did before Google, but I can.

I used something called AltaVista, which was the pioneer of web search and exploded in popularity in the late '90s. This is their 'simple-ish' homepage in 1996.

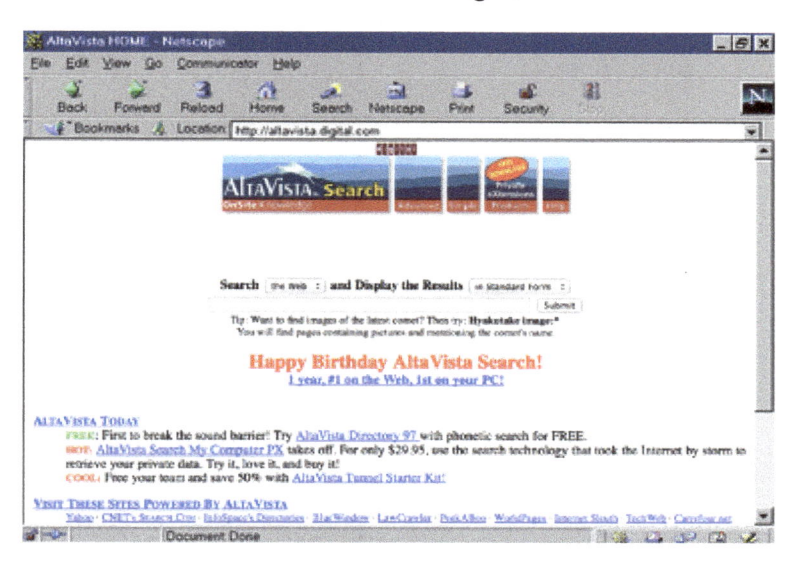

It did many of the things that Google did a few years later and had the advantage of the huge Digital Equipment Corporation (DEC) behind it. This meant they had easy access to computing power. They even (initially) had good relationships with Yahoo. After being acquired by Compaq they unfortunately (although seemingly logically) decided that they wanted to be an internet portal, like Yahoo, which led to the busy 1998 web page below.

Really Simple

Like all good 'almost' stories, this distraction created enough space for Google.

Google was all about simplicity and doing one thing well. Then they added in the additional search categories such as Video, Images, Maps etc. They missed the social media wave (although Gmail is huge), but I would argue that is because they were focused on their core offerings.

The rest is now history. Google does not publish its search statistics, but it is thought that they handle 2 trillion searches per year (at the time of writing) and generate revenue of about £160 billion across all businesses. All still with a one-box home page:

Processes

Process is where many of us spend a great deal of time. You might know them as work instructions, operating procedures, or form filling!

> ## RSX - Form Design Disasters
>
> How many times have you been asked to fill in a paper form (and sometimes now an electronic form online), with your name and address and contact details and there is simply not enough room? This can be so totally ludicrous that I have seen forms with a single line for name, phone number and address. At the opposite end of the spectrum, there are those with character-by-character slots, which make some sense but means that, again, there is often not enough space.
>
> For those thinking that this doesn't really matter, that's true it doesn't *really* matter compared to, say, solving world poverty, but it does matter a bit. Particularly if these forms are being filled out repeatedly, sometimes in huge volumes, and are meant to be for something useful or even important. What really matters is that, with just a bit of effort and sensibility, the designer could have produced something easy to use, which would make everyone's life easier.

Whatever the process type, it will mean following a set of steps, sometimes prompted, and sometimes learned. The processes that are Really Simple will stand a better chance of success than those that are complicated. Banks know this and, given that they are handling money, they have routines and procedures that keep track of and allow easy checking of credits and debits. The only real question is why they don't use simpler terminology, like 'payments' and 'receipts', or 'withdrawals' and 'deposits'.

RSX - Government Pandemic Payments

Governments rarely get wholesome praise. During the 2020 pandemic, the UK government created a scheme to fund 'furloughed' employees for up to 80% of their salary for six months, which was widely recognised as sensible and important.

Having created the furlough concept, however, the real problem was to find a way to process applications and payments that a) covered the wide range of circumstances, b) could cope with the sheer scale of processing, and c) provided a method of educating and checking those doing the applications.

From a standing start, the UK Government Digital Service (GDS) built the process and portal, leveraging existing infrastructure and standards. Most importantly, they also created a self-guided step-by-step process. This platform was up and running in 8 weeks and paid out approximately £40bn to over 1 million UK businesses, so they could keep paying their employees. While the furlough scheme had some issues, overall, it was a successful and simple scheme."

This particular success had been 10 years in the making and almost didn't happen because of internal wrangling over responsibility. GDS had been in place over successive governments and is broadly non-political (although potential risks to data privacy are still contentious).

Activities

Have you ever experienced that feeling of being in the groove with an activity and losing all track of time? Sometimes we just seem to be 'in the zone' and can get a lot done and feel incredibly good about it.

Rarely do we review the reasons why this is the case. Sometimes it is because the thing we are doing is super simple. Personally, I like raking leaves; what can go wrong? It's just me, a rake, and a lot of leaves!

As you move up the complication curve, you may require a set of inputs, along with a bit of uninterrupted time and access to the right tools. An example of this would be preparing a financial model for a project where you need to understand the costs and expected timings, together with a PC, a spreadsheet tool, and a few hours of quiet time.

To take this one step further, and to an extreme example, large capital projects like ship building have a phenomenal number of dependencies, all of which are made manageable through modern design, build and collaboration technologies. This is all supported by supply chains of individual companies that are specific experts in their fields. Whilst these might seem complicated in aggregate, they are only made possible by a huge number of simple elements or components.

Building a skyscraper is like this, with floor upon floor of repetitive building activity making something that seems difficult, surprisingly quick to construct.

Clifford Burroughs

Perhaps the most obvious example of the virtue of keeping activity simple is mass production. The ability to crank out domestic fridges at a cost that makes them available to almost everyone has transformed lives all over the world. Without the simple thinking that underpins production lines, combined with semi-skilled labour, many of the modern lifestyle benefits that we take for granted would not have been possible.

The key aspect to remember about any activity is the huge variation in output between simple and complicated, particularly when many people are involved.

Really Simple

4

Model: It's Multidimensional

Some things need less, others need more, and they interact

"Simplicity is an exact medium between too little and too much"
-
Sir Joshua Reynolds

As discussed in Chapter 2, complication is a mental reaction to whatever we are interacting with. This chapter details the things (dimensions) that commonly drive that mental reaction and provide a model to capture and create Really Simple outcomes.

Reductions and Increases

Reduction is the most common theme for those considering, theorising, or writing about simplicity.

In Edward De Bono's *Simplicity* (Bono's 2015), his fifth rule (of ten) is that "you need to challenge and discard existing elements". In *Insanely Simple* (Segall 2015), Ken Segall's

Chapter 3 is entitled 'Think Minimal'. In *Simply Effective* (Ashkenas 2009) by Ron Ashkenas, Chapter 3 covers 'Reducing Product Proliferation'. The Lean movement is also focused on removing non-value-adding waste.

John Maeda's, *The Laws of Simplicity* (Maeda 2006) starts with Law 1: Reduce. He then further elaborates this by providing types of reduction with an approach he labels 'SHE'. The S is for Shrink, the H is for Hide and the E is for Embody.

In his book *The Laws of Subtraction: 6 Simple Rules for Winning in the Age of Excess Everything* (May 2012), Matthew May details six specific ways of taking away complexity. A bit like Maeda, he does not claim to have all the answers, but he looks to expand on Maeda's main and all-encompassing rule detailed in the final chapter: "Simplicity is about subtracting the obvious and adding the meaningful". However, both books, good though they are, take a designer centric view, whereas RS is a more practical approach to work and life.

Really Simple

Al Gore had this famously overloaded office. He is probably able to block out the irrelevant to focus on the matters in hand, but for many this would be a mental nightmare. Just removing 'stuff' would be simplistic but may not have worked. He would need other meaningful things added to make him feel 'whole'.

In the context of RS, reduction is the critical need to thoughtfully remove unnecessary **building-blocks.** RS also embraces the need to add other **human-connection** characteristics that make a difference to simplicity.

The most obvious is establishing the right level of organisation that breaks something complicated into something understandable, like book categorisation in a library. Next you must make sure that your solutions are not entirely functional and create some kind emotional engagement. Finally, your solutions must consistently work as expected. All these characteristics are at the heart of the RED-CAR model detailed in the next section.

1. Resources
2. Elements
3. Dependencies
4. Comprehension
5. Appeal
6. Reliability

The RED-CAR Model

The RED-CAR model comprises six sub-dimensions:

Building Blocks	R	Resources
	E	Elements
	D	Dependencies
RS	C	Comprehension
	A	Appeal
Human Connection	R	Reliability

The first three need to be reduced and the second three need to be increased.

The reductions are focused on the **building-blocks**:

- **Resources** are independent of the number of components or dependencies and can be physical, time or cost.
- **Elements** are the constituent parts. These can be physical or soft for something like a process or system.
- **Dependencies** are necessary connections between the building blocks.

So, what do you need to increase?

Fundamentally you need to increase the **human-connection** characteristics which break down into comprehension, appeal, and reliability:

- **Comprehension** recognises the importance of usability. The two underlying attributes of this are intuitiveness and ease of learning.
- **Appeal** focuses on the importance of human desire or interest. This is probably the most challenging as it comprises usefulness, value, and aesthetics.
- **Reliability** calls out a much overlooked but very basic requirement. This breaks down into feature reliability and longevity reliability.

The words 'reduce' and 'increase' are used deliberately because Really Simple is an improvement journey. Reduction and increase are the direction of travel you will take to move to a Really Simple outcome. Less and more are the results of an RS initiative. As it is almost impossible to leap directly to an optimal outcome (which we will come back to later in the book), some stopping-off points will usually be necessary. Hence reduce and increase imply both a one-off initiative but leave the potential to improve more later on.

Measurable (Changing) Attributes

To create a practical and systematic approach to RS, it was important to understand the specific characteristics that needed attention. These are described here as attributes.

Three of the six dimensions have two or three attributes each. These attributes provide a structure to check both complication and simplicity. All the attributes are measurable, although some are more subjective than others.

There is also some interplay between dimensions. For example, fewer components might make something more reliable or easier to learn. The important point is that all ten attributes require dedicated attention.

Also, the first three are objective. Having the fewest number of sensible elements is not really subjective. The remaining seven are judged by their ubiquity; that is, the human connection with the majority.

Blocks: Elements

Seemingly, elements are the most obvious and easy attribute to see, change, and measure in any simplification exercise. The benefits of simplifying elements can include fewer products, fewer sub-assemblies, fewer suppliers, fewer steps in a process, etc.

Whilst this is something to address, you need to be careful of any unintended consequences, and you also need to be brave enough to drive the change. Removing steps from an investment approval process may make it quicker for those seeking approval but may reduce stakeholder agreement or understanding. It is often the case that stakeholders don't really use this involvement sensibly. Often the best solution is to provide some stakeholders with a 'speak now, or forever hold your peace' step in parallel to encourage speedy engagement with the process.

There are several ways to reduce elements, which will vary depending on the situation. The focus here is to reduce the count but this will also have a knock-on impact on the dependencies.

It may seem obvious, but you must agree what constitutes an element and therefore how to count it. For example, if you buy some elements of your solution from external suppliers, do you count those or just the elements you create from them. The decision is subjective and connected to another obvious point, meaning that you need to measure what matters, i.e., what will make a difference if it increases or decreases? Here you will need to be careful not to overlook what might look unimportant as it may be disguising an underlying problem. For example, you might be driving complication because of an underlying cheap resource.

The following topics describe the typical ways in which component reduction can happen.

Consolidation

Rarely can you make something completely redundant and therefore removable, without consequence, but sometimes you can create a single simplification as part of an overall RS initiative. For example, manuals for cars need to be produced in multiple languages. Over the years a simplification was to have one large manual with all the languages. This created fewer components but conversely created some resource (paper) waste because users didn't ever read the 9/10ths not in their language. This has changed again with the advent of Print-On-Demand technology, where it is so easy to print a few copies, single language versions are again sensible.

Instruction manuals are now moving online or being made available on the in-car screen systems, although this assumes that those are functioning! Ikea and others have taken this one step further to make its assembly manuals pictorial, meaning that no translation is even required.

Modularisation

Over the last 50 years modularisation has been a powerful enabler, used in lots of situations. It has been the backbone of growth in computer software and most visibly in global logistics.

The 40-foot shipping container is the universal building block of global commerce. If you accept that manufactured goods and produce need to be shipped around the world, the shipping container makes this economical and practical. Trains, ships, ports, and trucks are all designed to cope with this universally agreed standard, which has transformed global economic efficiency.

It allows for the repeating use of tried and tested components to achieve patterns of functionality that are much greater than they are individually.

Modularisation is also heavily used in manufacturing. Suppliers are often used to provide specialised elements of larger solutions, particularly where that specialisation has significant R&D overheads and can be 'bolted-in' with relative ease. Good examples are automatic gearboxes for cars and navigation systems for leisure boats, which are almost always 'bought-in' from dedicated manufacturers.

Organisation

The word 'organisation' is multi-layered and heavily used in day-to-day conversation. For some, it is a noun to describe a collection of people working together with some common aim. For others, it is describing how work flows through an organisation. For yet another group, it is the layout of a store cupboard. These are all relevant in an RS context, meaning good organisational thinking can lead to fewer elements being required. Consequently, a physical layout, ergonomics, or system flow may need to change, which also means fewer elements. With the right organisation, clicks, scrolls, movement (motion), process steps, communication channels, and cognitive strain can all be reduced.

Standardisation

You might not be able to reduce the overall elements of the user-experience, but you may be able to change the underlying elements of construction. This might mean one type of bolt for assembling some furniture, or if that is not possible maybe one size of Allen key.

Miniaturisation (Making Things Smaller)

This will often need a technical solution but sometimes just thinking about how to squeeze things into a smaller physical space will make a difference. The squeezing does need to be virtuous; it should not diminish the solution in other ways. Sometimes something will just be unnecessarily large and can be adjusted to the 'right-size'. I use miniaturisation all the time to ensure that I can present busy diagrammatic overviews, by simply using the 'Arial Narrow' font.

Singularisation

This means trying to reduce something down to one component. The best example is packaging, where often a mix of materials and colours can make it particularly complicated to both produce and recycle. Picking one material, even if it results in a slight increase in that input material, can make a big difference (particularly if that material is recyclable). As I open packaging, my heart either sings or sinks as I consider how easy or problematic it will be to recycle.

RSX – Automotive: Platforms, Engines, Switches

Over the last 30 years there have been countless mergers, acquisitions and divestments in the car making business. One of the key reasons is economies of scale; overhead can be shared, but more importantly manufacturing, suppliers, and supply chains can be consolidated and rationalised. In addition, the car manufacturers have been able to unify the foundations of every car.

The masters of this are Volkswagen who have four mass-market brands – Seat, Skoda, Volkswagen, and Audi (they also own Bentley, Bugatti, Porsche, and Lamborghini). Quite often, the same platform, engine and gearbox foundations will be used across all those brands. They will also leverage supplier relationships that specialise in, for example, air conditioning or seat manufacturing.

In one extreme example, Volkswagen brands were even sharing the same interior switches! Thankfully, they realised this was a mistake as the interior of the car is as important to defining the brand as the exterior, but they still probably buy their switches from only two or three manufacturers. In another initiative to reduce components, they have created significant sub-assemblies that can just be bolted on during manufacturing. This has required the manufacturers to create shared production plans with suppliers. This ensures that enough of the right sub-assemblies are made to the right specification, on the right day, and critically to the right quality, to ensure that rework is almost non-existent.

The benefits of this approach are literally countless. You achieve mass manufacturing benefits where you can apply automation and reduce costs either to improve value or profitability. In the supply chain, instead of there being 100 types of spark plug, maybe there are only ten, meaning inventory becomes more manageable. Therefore, the ability for designers, engineers, and technicians to keep track of the various possibilities in their heads also becomes manageable. Basically, some constraints take endless variety off the table whilst retaining enough to meet market needs. In the case where one of the mass-market brands does not fit your needs, there are now lots of options for smaller brands to operate using the latest niche production manufacturing techniques.

Blocks: Dependencies

Dependencies are often the silent killer in creating Really Simple initiatives. How often do you hear someone comment on something upstream that is stopping them from getting the important part of the job done? These problems are sometimes related to quality, sometimes to training or understanding, and sometimes just to timeliness.

Either way, in complex systems or organisations this problem can compound, having a debilitating effect on efficiency and, perhaps just as importantly, motivational energy.

It can also cause a negative 'network effect', meaning the more dependencies there are, the worse things get. These diagrams go some way to making the point.

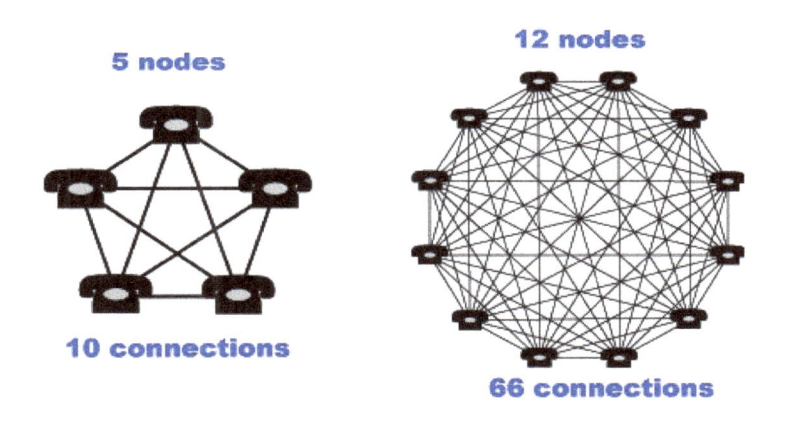

5 nodes

10 connections

12 nodes

66 connections

The implications on reliability, scalability, and communications are obvious and significant.

The following topics describe the typical ways in which dependency reduction can benefit the organisation.

Encapsulation

For a complex system to work, the independent components need to act independently and contribute to the whole. This usually means that there are a series of 'hand-off' interfaces, where responsibility is passed from one component to the next. However, inside the component itself just about any type of approach or solution is possible, provided it obeys the rules of the system. In the case of the internet, a set of hard-edge protocols keeps everything working, allowing for upgrades and improvements without stopping everything. Like modularisation, it allows for patterns of repeating functionality but also allows the implementation to be different behind the scenes.

Linearisation

Production lines all over the world lay out the various elements of manufacture and assembly to maximise simple flow. The accepted doctrine is that, where possible, straight-through processing is more productive, usually with less work-in-progress and a structured way of working.

RSX – Automotive: Context

The supply chain methodology known as 'Just-In-Time' uses the concept of 'through-the-wall fulfilment'.

For example, a specialist supplier of car seats can have a factory close to the main car production line (so they can metaphorically pass 'through the wall'). They produce specific seats to a specified order pattern (at the exact time it is needed - 'Just-In-Time'). Therefore, saving the need for this part of the production process on site. From the point of view of the car assembly line, it is relatively simple. The right seats turn up for the right cars (according to the production

> schedule) at the right point on the line for them to be installed. Of course, the planning and enabling systems need significant thought and organisation, but often this is not particularly challenging. The important point is that complexity is 'pushed away' from the expensive and critical process.

Blocks: Resources

Reducing components and dependencies might reduce the input resources; however, that might not be positive if the cost or the time required increase. The environmental impact also needs consideration.

An RS solution that has a reasonable shelf life is likely to have a lesser ecological impact than one that lasts for the metaphorical five minutes. However, regardless of shelf life, if the environmental impact of a product is recognisably significant, then this might also disqualify the item as an RS solution. The current hot topic is plastic, which traditionally might have enabled simplicity in many situations, but now the environmental impact must also be considered.

No special techniques are required here as measuring and trying to reduce input resources is a well-travelled road, but it is important to ensure that a solid approach to sustainability is in place.

> Whilst some plastics can be argued as worth the environmental impact, many are not. Single-use drinks bottles, plastic bags, and kids' plastic toys are at the top of the naughty list. Recycling and reuse play a part here, with

simplicity being a key enabler of facilitating change in disposal solutions. Whilst it remains far easier and more economic to unsustainably throw plastics away, progress will be limited.

Connection: Reliability – Features

Assuming that core functional needs are being met, then it is important that these functions continue to work for the intention that they were designed, meaning predictability. Whilst this feels very straightforward and obvious (it either works or it doesn't), it can be frustrating when something "sort-of" works.

Reliability is sometimes confused with quality, which is often used as a synonym for premium. Reliability should be evaluated distinctly from the features that define quality, suggesting that reliable can also be low cost.

You should also consider resilience (ability to recover) at the same time as reliability, so whilst a breakdown might be rare, the impact might be a disaster.

RSX – Antenna-Gate

Apple had made a significant breakthrough in the Smartphone market with the original iPhone and the next few generations (3G and 3GS). The iPhone 4 included some significant enhancements, but also had the unintended consequence of making the phone less reliable at making

> phone calls! This became known as 'antenna-gate', particularly as Apple wrestled with the PR that surrounded the problem.

Reliability feature testing will obviously vary greatly depending on the solution. The important point is to factor enough thinking and resource into assuring you meet your reliability objective.

Connection: Reliability – Longevity

RS outcomes need to have an appropriate shelf-life. The definition of 'appropriate' is fairly variable. For example, most people expect personal computers to last for around five years. However, buildings can, with the right kind of maintenance, last up to about 500 years. Most big software platforms tend to last about 15 years, although some heavily updated mainframe software systems are now 50 years old.

> As I write this, I am using a four-year-old Windows PC, a newish Monitor but also an additional old TV LCD screen that is 17 years old! I don't do this because I cannot afford a better newer monitor but rather because it works, and I want to minimise my e-waste.

It is hard to provide generic rules because of the huge variety of RS situations. The key point here is to have a realistic sense of the useful life that is expected and make sure that is factored into the design. Expected obsolescence or end-of-life also needs to be considered from the outset. If the RS challenge is commercial, then this will be important in understanding the total economic impact and any resilience implications.

Connection: Appealing – Useful

This is the old East German Trabant interior. This grainy picture does not quite show it, but this car does not have a petrol gauge. I know the designer could make the case that the user knows how far they have travelled and therefore could work out when they will run out of fuel, however, by most measures, this is missing basic functionality.

Cars these days estimate the distance you can travel until you need to fill up and prompt you when you are down to your last 50 miles. Whilst these indicators are commonplace now, there was a time when this innovation provided really appealing functionality.

Connection: Appealing – Valued

Whilst useful is necessary, valued is critical. Useful means that whatever you are simplifying must fulfil some sort of purpose. Being valued means that you would be willing to trade something else to have this. Usually it will be money, but the valuation might fluctuate quite widely depending on the situation. For example, many overheads are necessary inside an organisation, but are still costs which customers will not recognise.

In a world largely regulated by economics, you will need to make decisions about costs and pricing. An RS approach will usually, but not always, imply a lower relative cost solution, meaning you are likely to have cost competitiveness. Depending on the nature of your organisation, you may want to maximise profits or, in the case of healthcare items, promote take-up.

Another increasingly recognised form of value is that of saved time or improved well-being. Both can now be measured and might often be facilitated by the success of one of the other attributes.

Connection: Appealing – Aesthetic

For something to qualify as Really Simple, it needs to have a pleasing form. That might be how it sounds, feels, looks, or smells.

This (below) is the Fiat Multipla. It is considered probably the ugliest car ever. It could qualify as RS in just about every other aspect, but it would be undone by how it looked. For the few of you looking at this thinking it does not look too bad, that's not the point. The point is that simplicity is not in the eye of the beholder but rather needs to be ubiquitous.

Really Simple

Ubiquitous aesthetic is quite a challenge. Seemingly the most successful way to achieve this is to rely on traditional aesthetic rules, which unsurprisingly include the use of simplicity: symmetry, balance, and rightsizing.

Connection: Comprehension – Intuitive

It used to be very hard to get computers to do what we needed them to do. They were big machines, kept in special air-conditioned rooms and managed by computer engineers.

Computing has obviously come a very long way in the last 50 years. Even compared to the computing power of the Apollo 11 mission, which landed a man on the moon, a modern Smartphone has 1 million times the memory and 100,000 times the processing power. This has made the software incredibly rich and, for phones and tablets in particular, means that kids can use the technology from a very young age.

Whether a tablet or a smartphone is a good example of intuitive use is still a matter of some debate, but their pervasive and fast adoption across just about every geography and demographic implies that it is highly likely.

Contrast that with the remote controls for your television or the buttons on your washing machine, both of which will usually need a manual, at least whilst you familiarise yourself with the basics. And this is despite the attempts of various designers over the years to use icons and some common standards. They are not terrible, but they are also not great or even remotely good.

Making things pervasively intuitive is about visual clues and logical structure. Logic in this situation is the norm; mental models and metaphors that people have developed over their lives, although they may be a bit illogical, represent their perception.

Connection: Comprehension – Learnable

Unfortunately, not everything comes ready assembled or fully configured, meaning some instruction will be necessary. The method and quality of instruction has been transformed over the last 15 years, facilitated mainly by the internet, particularly through video clips, and delivered by YouTube as the preeminent platform.

However, it is still the case that most instructions come in the form of pamphlets or manuals, even if they are available online. Whilst this will change over time, in the worst cases these can be infuriating, but in most cases just a bit frustrating, especially as you try to find the right bit in your language!

Judging whether something is easy to learn is not easy for the designer or even maker, as they are often so familiar with the solution, so testing with a meaningful sample of users will almost always be necessary. You may even need to add some elements here to allow for different types of learning - visual, doing, classroom, etc.

Really Simple

The RED-CAR Template

This is the Really Simple RED-CAR template, a useful tool to assess the nature of your simplicity or complication.

This template provides a way of prompting and recording the thinking behind all the attributes that make up the RED-CAR model. You can use it to either document the as-is situation or the to-be future situation, or even a hybrid of both.

REALLY SIMPLE **REDCAR** TEMPLATE			
Date:		Topic:	
Building Blocks		Resources	• • •
		Elements	• • •
		Dependencies	• • •
Human Connection	**Comprehension**	Intuitive	• • •
		Learnable	• • •
	Appeal	Useful	• • •
		Valued	• • •
		Aesthetic	• • •
	Reliability	Features	• • •
		Longevity	• • •

5

Mindset: It's Psychological

The 5 i's

> *"Everything should be made as simple as possible, but not simpler"*
>
> \- Albert Einstein

It is critical to understand the nature of complication through the RED-CAR (dimensional) model detailed in Chapter 4. However, it is just as important to understand the significance of mindset. Without the right mental attitude, the RED-CAR model will only provide a foundation, not a Really Simple outcome. These 5 i's will help propel RS theory into a RS reality. As with the RED-CAR model, the 5 i's need to be calibrated appropriately for the challenge at hand, the context, and the individuals involved.

i1: Intention (or Highly Motivated)

It might seem obvious, but you must *really, really* want to create something Really Simple. That is because, as we have already confirmed, it is not easy. You will need to understand the potential benefits and really believe that it is worth the uphill climb to get over the frustrations and complicated, often subjective, decision making.

Over the years, I have seen many a manager, colleague or friend trying to deliver a so-called simplicity initiative, but lose momentum, motivation and sometimes ideas. There are five key reasons for this:

1. The truth is that in life and work (in the West in particular) we are so conditioned to use prioritisation (or Pareto's 80/20 thinking) that compromise is considered a management virtue. This is what a colleague termed "that'll do" thinking. (It should be said that some nations/cultures are better at staying the course than others and this is not just limited to simplicity initiatives.)

2. In life and work, there are constant streams of competing pressures and things to manage. This can drive significant distraction from delivering Really Simple initiatives. Having made some improvement, you might be tempted to just move onto something that seems more pressing, and that is still often a sensible idea in many situations.

3. Sometimes it is not clear where to start, as the challenges can seem so interlinked and intractable. Some people wade in using destruction as a conscious tactic before they start to rebuild and create something new. This approach uses generally accepted management machismo as a vehicle of change. It is a personal choice, but in my experience, unstructured destruction creates mistrust and credibility issues that outweigh the limited benefits. Change and renewal are inevitable, and part of the circle of life and

work, but rarely is total destruction the way to create a platform for the long term.

4. Any individual's willpower is not an unlimited resource and staying the course is part of the challenge.

5. Determination to make things Really Simple will not necessarily make you popular. In fact, it may cause the opposite effect and you are cast as the bad guy. If you "get it" and others do not, then it will be a bit of an uphill struggle. (Obviously, giving any sceptics a copy of this book is a great answer!)

With so many potential barriers, a truth is that top-down understanding and leadership is a critical enabler. This is the business cliché that trumps almost all other business clichés. If I was given £1 every time I have heard a colleague, commentator, or leader say that executive ownership is the key enabler, I would be rich. That said, clichés are popular for a reason. In this case, the leaders must believe in simplicity as an enabler, to provide the right context to succeed and facilitate the 5i's.

In *Think Simple* (Segall 2016), Ken Segall provides many examples of how leaders can play a role in making simplicity happen for the good of all those involved. Whilst a high proportion of the examples relate to Apple (Segall was a long-time collaborator with Steve Jobs), he provides thought-provoking challenges on how to counteract anything complicated with simplicity.

Segall has a background in the creative world, and it shines through in his strong preference for as little bureaucracy as possible. His book prompts and answers many questions about simplicity, providing ideas on both reducing building blocks and increasing human connection.

RSX - Elite Sport and Seeking Marginal Gains

Over the last 15 years, British Cycling has been transformed to the point whereby both Olympic and tour successes are now regular. This success is largely attributed to Sir David Brailsford, who popularised the concept of 'marginal gains'. His approach was disturbingly simple. Just find many small (1%) improvements over time and the aggregate could be world-beating. It was not new, because Lean effectively advocates the same thing through continuous improvement, but it did bring a truly simple method and focus to a sport that had traditionally focused on just the rider and the bike. It probably should also be said that ample funding was also a major enabler, allowing Brailsford to chase down each percentage point without penny-pinching.

Clifford Burroughs

i2: Intensity (or Sustained Tenacity)

> *"We choose to go to the moon in this decade and do the other things, not because they are easy, but because they are hard"*
>
> \-
>
> JFK

In my experience and observation, and in folk law, nothing worth doing is easy. Almost no one will set out on their simplicity journey with an ambition as lofty as going to the moon, but in some cases the effort required may not feel much different. This intensity will need to be sustained. Moon-shot endeavours are rarely overnight sensations and are often the result of years of 'hidden' learning, experimentation, and application.

To get the best from intensity, the environment needs to be right. It is no good toiling on the wrong things, so you need to have the right kind of intensity. In his book *The Way We're Working Isn't Working* (T. Schwartz 2010), Tony Schwartz establishes that humans are designed to work intensely, followed by periods of rest. In other words, humans are not like computers that can work and work without getting tired. Worse still, tiredness has a vicious circle effect, reducing sleep and impacting brain power. Characters from history have intuitively understood this with many of our most celebrated thinkers and inventors taking an afternoon nap (or other forms of rest).

Really Simple

Aristotle, Albert Einstein, Thomas Edison, Leonard da Vinci, Salvador Dali, Winston Churchill, John F. Kennedy, and even Margaret Thatcher were all famous fans of power naps.

In his book *The Gathering Storm* (Churchill 1948), Winston Churchill stated: "Nature has not intended mankind to work from eight in the morning until midnight without that refreshment of blessed oblivion which, even if it only lasts twenty minutes, is sufficient to renew all the vital forces… Don't think you will be doing less work because you sleep during the day. That's a foolish notion held by people who have no imaginations. You will be able to accomplish more."

It has been said that Churchill inspired many leaders that followed him to take up this practice. However, apparently where some had been proud of this method, others thought it seemed unreasonable. Ronald Regan apparently did not want people thinking he was lazy for having daytime naps.

"The reasonable man adapts himself to the world; the unreasonable one persists in trying to adapt the world to himself. Therefore, all progress depends on the unreasonable man [and woman]"
-
George Bernard Shaw

> ## RSX - Packaging Apples
>
> In the introduction to his book *Insanely Simple* (Segall 2015), Ken Segall talks about "The Simple Stick". The example he uses quotes a member of staff returning from a meeting with Steve Jobs on the packaging design for an item. The quote was that Steve had hit them "with the simple stick". The team had been developing two packaging solutions, for ostensibly the same product, and Jobs said that one product should mean one packaging solution. He recognised that focus and efficiency was the winner and customers would indirectly benefit.

Sometimes to create intensity you need to overwhelm a problem with resources and focus. This might mean creating a dedicated team, with specific capabilities, in a dedicated location with a different set of management rules.

> ## RSX - Cracking Enigma
>
> At first sight, this might not seem like a simplicity example, but the thinking was: firstly, get the brightest mathematicians, engineers, and linguists, and locate them all in the same (secure) place (Bletchley Park). Secondly, give them a specific 'moon-shot' challenge with some organisational discipline. The intention, iterations, intelligence and investment, combined with intensity, allowed a rapid breakthrough, where messages were being broken at the very beginning of the war. This meant that despite further German innovations, they were able to keep deciphering messages throughout most of the war and keep their

success a secret. Finally, they relied on the everyday habits of the average German operator who, despite having a fantastic encryption device, might include predictable messages within the text, substantially reducing the potential computational challenge. For example, many messages would start with 'To' and sign-off with 'Heil Hitler' with specific operators having specific habits. Without these 'short-circuits' the rudimentary mechanical 'Bombe' machines would have had to work all day solving a message that would in that time often become too out-of-date to be useful.

i3: Iterations (or Keep Improving)

It seems to take at least three versions of most things to get them right, although this is often forgotten in the afterglow of success.

RSX - Windows Everywhere

Windows for the PC is the ultimate example of try, try, and try again, although there are many other examples in software. Back in the '80s, Microsoft was making a splash in non-graphical software like DOS and Basic. However, like Apple and VisiCorp (who invented the spreadsheet), they had also seen the potential of graphical user interfaces at Xerox's research labs. They developed and shipped Windows 1.0 in Nov 1985, but in truth, this version had little capability. It did, however, signal the way in which the market could develop by providing software developers with a foundation upon which to build their offerings. The power of

> PCs increased considerably in the second half of the '80s and Version 2.0 was released in 1987 and version 3.0 in 1990. Whilst both were step changes, they still did not achieve a big market presence. Finally, in 1992 they shipped 3.1 which took over the marketplace (in part because it worked), because it had good software like Excel and Word, and finally it had good compatibility with so many hardware providers. (Another non-product-based reason was that businesses had started to back Microsoft as distinct from IBM and Apple, which had made little progress providing mainstream hardware and software.)

Without the virtuous constraint of making something Really Simple, rarely will something start and stay simple. This is because a seemingly good enough solution can be achieved with a few iterations, and then other pressures kick in requiring us to move onto something else.

The term 'good enough' is the bane of modern professional management. It's a biproduct of the MBA generation, where statistical averaging drove mediocre business decisions. In a world that is in growth, this can work; however, in a world of stagnation it does not.

Designing a house, a consumer product, or a software system requires the same basic iterative thinking, but most importantly 'good enough' should not be the criteria for stopping.

i4: Intelligence (or Good Thinking)

As we have established, making things Really Simple is not really easy. Making the decisions about what is and what is not RS is much harder than most appreciate.

I have explained how intent, intensity and iterations are important, and whilst those i's are not obvious, most readers will consider them useful and recognisable concepts which could impact in most aspects of their life.

The topic of intelligence, however, is far more sensitive. The opposite of intelligence is unintelligent, and most people would think of themselves as intelligent, and that is part of the problem; believing that because you are intelligent, that you will be able to create and apply simplicity. This section is about recognising that this can be an illusion, by highlighting both potential traps and less obvious opportunities.

Recognise

Simplicity Is Mostly in the Eye of the User

Firstly, it is important to remember that an RS solution to any challenge must be simple for the user. It is the job of the designer to ensure that the outcome or process reflects RS criteria. Whilst the designer can be an expert in their field, they are unlikely to be experts in simplicity. It is critical to ensure that designs are reviewed against the RED-CAR model from the perspective of the user.

For many of us, putting ourselves in the shoes of the user can be problematic because we will usually be part-time or occasional designers, with day jobs, which means the solutions may be different to a designer trained in simplicity.

Your situation may be very specific, with tens, hundreds, or thousands of users. You need to actively consider the likely impact of a poor solution in both the short term (initial reaction or moment of truth) and the long term.

> Professional designers (and architects) are not perfect of course and can often prioritise form over function, meaning practicality is sacrificed.

Sometimes it is Blindingly Obvious

On occasion, some things that have had significant thought, and are relatively successful, still fail because of something relatively basic. The most obvious is reliability, such as when computer-based solutions are either glitchy or slow. You can focus on fixing these something specific and leave everything else alone. These fixes may not be that easy, but you at least know where to target your resources and the outcome is measurable and tangible.

Intuitively Seeing the Answer

It is interesting to observe that those who are currently successful at simplification often have a hard time explaining exactly how they achieved it. They will have half an idea and might describe this as intuitive and most of the time that will be partly true. Breakthroughs in simplicity do, however, need a level of conscious and structured thought, which may seem intuitive but usually have been strengthened with years of experience. This is a bit like a chess master who can magic up incredible moves out of thin air. These grandmasters have spent years in intense practice to wire their brains to the point that they literally think differently.

Old Fashioned Experience and Intuition

Whilst most of this topic (intelligence) is focused on objective intentional thought, we should spare a thought for experience and intuition. Experience can drive decision making that is hard to rationalise but just seems right. In Malcolm Gladwell's book *Blink* (Gladwell 2005), there are numerous examples of first thought correctness and what is known as 'thin slicing'. This thinking recognises the diminishing benefit of experts making decisions based on limited observations or data points, meaning they can make pretty good decisions quite quickly. An example of this is the instinct of experienced fire-fighters, who can process a number of parameters as they enter a burning building and can seemingly instantly know whether it is safe or not.

Hopefully, your simplicity challenges are not as acute as this example, but you should be conscious that your first thoughts will often be of significant value, assuming of course that you are an expert, and the data points are a valid sample. Without the pressure of flames flickering around you, you can then invest some time ensuring that you avoid some of the 'gotchas' below.

Gotchas

Simplistic Thinking

Sometimes the problem is relatively straight forward or urgent and the first thought is the right one. There are times for simple answers to simple questions. However, as suggested before, this can cause some people to declare victory too soon. You need to counteract this by asking if all the RED-CAR components have been carefully reviewed and addressed.

The Obvious Option Is Not Always the Right One

The right thinking mostly means picking the right option, but sometimes it is easy to be carried along by trends or groupthink, making the case one way or another. Very often, this will mean pressures to add something such as a new flavour, a new data check, or a new pack size.

> If you are facing pressure for increasing complication, it is no good shying away from it. This should not mean trench warfare, as you should be able to make the case with data and logic, but it may require more time than you might be hoping to spend. In this example below, I was conflicted because my instinct said to keep the project simple, but the sponsors of the project wanted something that looked interesting on their CV.

RSX - Project ONE (One Northern Europe ERP*)
*ERP = Enterprise Resource Planning software

I was lucky enough to have only a peripheral role with a big ERP (SAP) project in 1999/2000. I say lucky because I had a young baby, and the project was a bit of a death march for many of those involved. It was a re-engineering project with all the 5i's and had simplification and efficiency at its heart but struggled under the aggregate scale of the change. It was a very tough endeavour, and many swore never to do it a second time. However, because I had only been slightly involved, I jumped at the opportunity to do a smaller but similarly complicated project.

I had been assigned to a region that had a hotch-potch of systems following several acquisitions. The region was not making any meaningful profit and, more importantly for the private equity owners, it was cash negative and effectively being 'propped up' by other businesses. With limited potential for top line (sales) growth, the business needed focus and cost reduction, particularly reducing headcount, manufacturing, and procurement. Following the initial organisational intensity, which focused on rationalisation, it became clear that the underlying processes and systems were getting in the way of further rationalisation and efficiency. So, the IT team swung into action looking for other, better solutions.

The most obvious solution was to leverage an existing significant investment and capability in SAP, which we used in other territories. That would have made sense in the longer term. However, it would have been a very costly project, so we decided to leverage another existing investment in a less complex ERP (JDEdwards), which had less capability but was already being used in parts of the company.

This was when we made our first really big mistake. We had a fundamental challenge to convince the business that the change was worth the pain, and we needed to focus on business processes. It was felt by a few influential senior managers that we could not sell the idea of a 'new' system that was based on 'old' AS400 green screen technology. We were in a world that was almost all Windows and had started to move to web browser interfaces. The problem was that all our internal experience was on the older AS400 platform and even more profoundly there was very little actual external experience of the new client server version we had chosen to deploy.

It is hard to quantify the levels upon levels of complication that this decision drove. It delayed the project by about five months whilst we figured out how to get the platform right(ish) and then for the duration of the project we were regularly butting up against basic and unexpected technical issues.

The truth was the AS400 "devil we knew" would have given us all the process and rationalisation benefits, for less cost, less time, and less stress. We could have focused on the change management and process improvement. Yes, we would have been operating with some constraints, but they were known, and we could have done the client-server 'upgrade' afterwards. Instead, we were often hunkered down just trying to get the basic tech to work.

> Fortunately, the organisation had maturity and experience and knew this was going to be a challenging project. They stood behind the decision and made it work. I hasten to add, I never blamed the supplier (Oracle) because they never particularly gave us a bum steer, I blame myself for not being strong enough with the organisation of the consequences. We did obviously re-calibrate the plan when we switched our approach, but that plan did not account for the level of risk we had injected.

It's for the User, Stupid

As we have already discussed, designers might be part-time, subject matter experts that inadvertently design a solution for their needs. These experts are critical, but their motivations and minds-eye picture of success might be totally different from the user base.

I once oversaw an employee expenses solution that was created by the tax department. The tax reporting was world-class, and the tax department received significant professional kudos as a result. Unfortunately, the thousand people that used the app every month to submit their expenses loathed it!

Don't Design for the Tail

Most products, services and situations have a few aspects that are heavily or extensively used. However, the overall list may be quite long and one of the challenges is how to manage the wide set of needs without burying the main aspects. To some extent deciding which requirements are necessary is driven by 80/20 [Pareto] thinking, whereby you need to judge the benefit delivered, versus the resource required of the lesser-used elements.

The important thing is not to feel obliged to include every aspect of a perceived requirement. The best example of this is to hide the occasional features or processes in some way. This can be taken literally, with features hidden behind a physical or virtual door. The other common way is to provide menu systems that allow access to the regularly used elements at the very top, and those less used are hidden. This does mean that unused features will *remain* unused features, so you need to decide if that makes sense or if it would benefit from a change. Well-designed app menus and navigation is incredibly important, particularly with the small(ish) screen sizes of mobile phones.

Another strategy is to focus on the core and then 'add more'. In this situation, it is common to refer to a Minimum Viable Product (MVP). This allows for further extensions, additions, and modifications as the solution or product is used. This strongly correlates with the need for iterations and encourages clear thought of the benefits for one or a group of users.

Adoption Curves and Inflated Expectations (Misplaced Optimism)

Recognising the reality of adoption curves is also critical. It is rare for anything substantial or significant to be an overnight sensation. It might sometimes look like that, but for every success there are often several failed attempts or competitors that have now fallen by the wayside. An iterative approach embraces this reality, but more importantly, your expectations need to. High expectations can encourage early momentum and encourage support, but often those 'growth curves' can be entirely unrealistic.

> ## RSX - The Stockdale Paradox
>
> We are not innately aware of our heuristics (mental shortcuts) and cognitive biases because they are 'built-in'. It often takes an expert doing research to both identify them and consider the potential implications. The so-called 'optimism bias' applies to about 80% of the global population, regardless of gender. It means that in any given situation you will naturally expect things to be better than statistically would be sensible.
>
> Having optimism is a good thing in a lot of circumstances, resulting in either a positive or, at worst, neutral impact. In fact, optimism seems to be a feature of nature, where all kinds of wildlife assume the best most of the time and reacts instinctually to danger through fight or flight. Perhaps most importantly, it is so intrinsic that it is extremely hard to switch it off biologically and is noticeable in those suffering from mild or severe depression.
>
> The implication is that this needs to be managed. The Stockdale Paradox was made popular in the Jim Collins book, *Good to Great* (Collins 1975). It is named after James Stockdale, a Vice-Presidential candidate, naval officer, and Vietnam Prisoner of War (POW). The paradox is based on the contradiction of the words 'realistic' and 'optimism'.
>
> Stockdale was held captive as a POW for over seven years, experiencing torture and horrendous conditions. Stockdale recognised that you needed to embrace and accept the

> terrible realities and combine it with a healthy dose of optimism. Stockdale summarised this by saying: "You must never confuse faith that you will prevail in the end — which you can never afford to lose — with the discipline to confront the most brutal facts of your current reality, whatever they might be." His observation was that those with unbridled optimism usually did not survive.

This example has two significant implications for the RS approach. Firstly, if you are not ready to stay the course, you will be undone either by yourself or others. The second, is that you might need to manage the expectations of others to ensure that you manage to move the project forward to a second or third iteration.

Don't Confuse Perceived Speed with True Speed

After a substantial career in technology development, I have had my fair share of project overruns and software application issues, and poor project setup or monitoring was rarely the problem. Most of them were caused by overly ambitious schedules, either driven by arbitrary (i.e., very ill considered) deadlines or totally ignored realities, the key one being complication.

> The second most common reason for project failure is access to, or the availability of, the right resources. However, you need to always remember the adage that nine pregnant women cannot make a baby in one month. i.e., Schedule is King!

Really Simple

Most software has a set of features that are usually well understood by the management but only make-up a fraction of the code that needs to be developed to make it usable. This means that in classic iceberg style, 7/8ths of the work is not understood, it is 'below the water'. To make matters worse, for many of the projects I was involved in, it was often possible to create good 'mock-ups' that were critical for achieving clarity but would give a very false view of effort required to get the software into a market ready state. This leads to unrealistic optimism about what is possible.

We have referred to iterations, realism, and minimum viable products, but the ability to understand the difference between a quick superficial win and deeply embedded improvement is a key Really Simple enabler. Getting a great outcome is rarely a frenzied dash for the line, and thoughtful and intense (but not slow and steady) will win the race.

Appreciating true speed also helps shape the design of any solution. Keep in mind that users of the solution might not fully appreciate the difference and you need to either accept that or find a way to broaden their understanding. For example, a quick wash function on a dishwasher may not include the drying function of the long wash, so it seems shorter but in practice takes longer end-to-end. Another common example is where a team is working exceptionally long hours and weekends, but the quality and productivity of their work drops so significantly, they might as well not work the extra hours. Sometimes as a manager the most important act is to send your team home to get some proper rest.

Clifford Burroughs

Don't Let Your Eyes Deceive You [WYSIATI]

A core challenge is what Daniel Kahneman in *Thinking Fast and Slow* (Kahnerman 2012) describes as 'What You See is All There Is' (WYSIATI). This phenomenon means that when presented with small amounts of specific information our brains easily extrapolate the data to a fuller description, making substantial judgements about the context they are presented with.

When this happens in relation to simplification the risk is twofold. Firstly, that you will use this power to legitimise a suboptimal solution for both yourself and others around you, preventing further iterations and intensity. Secondly, you may just come up with a bad solution. Uncharitably, this could be described as a simplistic response, but it is how we mostly operate in new situations.

Beware the Heuristics and Biases

Good thinking is mostly about reaching good decisions. (Otherwise, it is usually about moving the body of knowledge forward.) Our brains are made up of two distinct ways of working: instinctual and conscious. Interestingly, instinctual thought is very effortless and quick, whereas conscious thought is tiring and slow. Conscious thought is also far from perfect when it comes to anything slightly complicated. To help with complications, our brains develop mental shortcuts, which are often helpful but can sometimes be wrong. These shortcuts are known as heuristics.

Kahneman's 'WYSIATI' approach illustrates one of the most important heuristics and that is 'availability'. This determines that people overestimate the importance of the information available to them. For example, from listening to the News you might conclude that the streets are not safe, when the actual statistics (of which you are unaware) say that they are. That is, you place disproportionate importance on the information you have to hand, e.g., news reports, rather than the important (yet unseen) underlying facts.

> Obviously, the ideal situation is balanced journalism but even then, it is still incumbent on us all to make enough time to understand important and complex matters.

The impact of these heuristic shortcuts is so significant, they have been studied and are known as **cognitive biases**. They describe common and repeating acts of irrational thought based on our inbuilt mental mechanics.

Clifford Burroughs

The significant biases which have a dramatic impact on our perceptions and particularly on simplicity are listed here:

Bias	Description
Anchoring	Over reliance on the first piece of information to set a range.
Confirmation	We tend to only 'tune-in' to opinions or facts that fit our preconceptions.
Bandwagon	Otherwise known as groupthink, where a group of people maintain harmony or conformity by reaching a sub-optimal outcome.
Recency	The tendency to over-value the most recent information.
Stereotyping	Expecting that someone (or a group) will behave in a certain way without truly knowing the reality.
Overconfidence	Over-belief in your own ability and rationale.
Outcome	Believing the outcome of something is correlated to a specific input.
Information	Wanting more information that will not improve the quality of the decision.
Clustering	Seeing patterns where there are none.

Deliberate thinking is required for RS to not let these cognitive biases effect your decision making.

> Experts will obviously develop experience and knowledge that can counteract the impact of biases. However, in day-to-day life you will see countless examples where these biases drive crazy outcomes but are completely accepted solutions. They are so entrenched in life that people will often be unable to accept there is a problem. Don't let awareness of these biases drive *you* crazy and accept you can only

> influence change in certain circumstances. Be aware that when you are becoming frustrated with a situation, it will often be because your sensitivity to bias will be at odds with the context you are in.

Emphasise

Lean Thinking Is a Good Start

I was lucky enough to spend a couple of years working with a 'lean' guru, and whilst he said many profound things, there were a few points that stuck with me. More importantly, this expert (and others) focused on Lean culture, knowing that the right daily behaviours make the difference. As a reminder, Lean thinking is fundamentally about removing all waste (from any system) so that you maximise value. This is where the reduction of elements in a RED-CAR analysis would apply (see Chapter 4).

Lean thinking is not (intellectually) difficult, it is mainly about using common sense. The real problem is that people often think the simple things are almost not worth bothering with, that maybe they are beneath them. For example, Lean thinking requires keeping things really clean to see any potential defects and by maintaining an environment you maintain standards. In effect, it is the equivalent of a soldier maintaining a rifle at the very highest standard, so it doesn't let them down at the critical moment. Cleaning in many walks of life is considered menial rather than critical; cleaning is often left to the cleaners. (Deep cleaning may require more advanced expertise but is only necessary periodically.)

The second connected behavioural habit that the Lean guru promoted was that "there should be a place for everything, and everything should be in its place". His example was domestic kitchens, run by a professional homemaker. These, homemakers (usually women) were models of Lean behaviour, expertly managing the household and minimising any waste of food, time, and movement. The high standards were achieved through extreme cleanliness, stock rotation, and hyper-organisation. They often learnt from their own mothers, who may have also had part-time jobs. They would be tasked with the bulk of daily parenting duties and had husbands that might be working long hours, but still got everything done by thinking about and needing to manage their efficiency.

Cognitive Ease

In their book *Algorithms to Live By* (Christian and Griffiths 2017), Brian Christian and Tom Griffiths explore various decision challenges that have been partly addressed by the computer science community. Throughout the book, they tackle the various decision archetypes but conclude by suggesting that instead of being forced to deal with complicated choices, everyone should try to practice computational kindness. This means that we need to consider the brainpower required for someone else to solve a problem, even if this means restricting some of the potential options. A typical example is when you are trying to schedule a meeting; being flexible gives lots of options but is harder than a yes/no answer to a few specific options.

To some extent, we all intuitively do this when we give people instructions to a destination, not necessarily providing the shortest, most complicated route, but rather the one that is more straightforward to navigate. We also see the other side of this equation when someone writes to us suggesting a meeting but not providing a specific time. Faced with too many options we often end up deferring a response, rather than just saying yes or no to a more specific suggestion.

Really Simple encourages 'cognitive ease'. Any RS solution needs to feel like there is no mental friction. This is recognised and documented by Daniel Kahneman in *Thinking, Fast and Slow* (Kahnerman 2012). His research shows that choosing the right colours, fonts, and visual structure are significant drivers of cognitive stress. Whilst he doesn't look at the simplicity implications explicitly, cognitive load can significantly influence the quality of any experience.

Deliberate Empathy

Many of the models developed for design-thinking start with empathy. Whilst most of us will have some degree of human empathy, to be a good designer means 'getting inside the head' of the user in a structured way. Fortunately, there is a recognised tool for doing this. The Empathy Map was originally developed by the consultancy firm XPLANE but has evolved over the years.

The Empathy Map requires the designer to understand how their target audience 'relates' to the challenge being tackled. They need to investigate the four quadrants:

- how they think and feel;
- what they say and do;
- what they see;
- what they hear.

In addition, an Empathy Map will capture the current pains and potential gains of moving from the current to an improved situation.

Fostering Creativity

> *"Creativity is contagious, pass it on"*
>
> \- Albert Einstein

Whilst this book is orientated around systems thinking, a sizable chunk of time will need to be devoted to creative/innovative endeavours.

There is a sizable body of work focused on the subject with lots of competing perspectives, so here we will simply say that you need to be in the right frame of mind to be creative, and there are a few recognised methods that make this easier. (By the way, these are good suggestions of things to do full-stop, not just to stimulate creativity.)

Stimulating Creativity

Change Scenery	Take your brain away from the place that you work regularly. That might mean grabbing a laptop or some paperwork and heading to a cafe.
Change Routine	Habits can be very powerful at getting things done but if they become rituals then they become stifling.
Take Breaks	You are not a computer that can just run and run. You have natural ups and downs in the day which you should seek to understand and manage around.
Keep Learning	Reading and research should ideally take about 5% of your time. It stimulates ideas and keeps the brain maintaining its critical thinking capability.
Enough Sleep	Broadly to have good brain function you need to have between seven and nine hours of good sleep each night.
Some Exercise	You don't have to be an Olympic athlete but getting your heart pumping and strengthening your muscles will improve your general wellbeing.
Good Diet	You need to (literally) feed your brain with the right things. Less meat and sugar. More vegetables and fish.
Walking	It seemingly clears the head.
Capture Thoughts	Ideas come to us at all sorts of odd moments. I have lost count of the times when great thinking is lost because I had no system of capture. Carry a tiny notebook and pen or use your phone to take voice memos or notes.

The above list is mainly focused on individual creativity. There are obviously several situations where group creativity is critical. In those situations, finding ways of stimulating creativity will be necessary, but usually the power of a group will create energy and interactions. In truth, the bigger problem will often be how to corral outputs into something useful.

Ergonomics

Ergonomics is the way humans interact with the world (and the systems) around us. A more formal definition is:

> "Ergonomics is the scientific discipline concerned with the understanding of interactions among humans and other elements of a system, and the profession that applies theory, principles, data and methods to design in order to optimise human well-being and overall system performance."
> (International Ergonomics Association)

Ensuring safety and general wellbeing is at the heart of ergonomics and should be at the heart of your RS thinking.

For example, we take road signage for granted but it is a critical system for the humans trying to navigate complicated junctions.

To achieve best practice design, Ergonomists use the data and techniques of three core concerns:

Physical	Human anatomical, anthropometric, physiological, and biomechanical characteristics.
Cognitive	Mental processes such as perception, memory, reasoning, and motor response.
Organisational	The optimisation of sociotechnical systems, including their organisational structures, policies, and processes.

Really Simple

Whilst this book does not attempt to cover these topics, it is critical that you identify what might be important to your challenge and make sure that you include the right thinking. Some of this might be obvious and easy to solve but as described before, 'the easy' is easy to overlook.

Inquisition: '5 Whys' - What Really Matters

One of the classic tools of problem-analysis is using '5 Whys'. This tool is built on the assumption that when you are trying to get to the root cause of a problem you usually need to ask 'why' about five times to get to the answer.

Here is an example of '5 Whys', which looks at why a customer delivery was late:

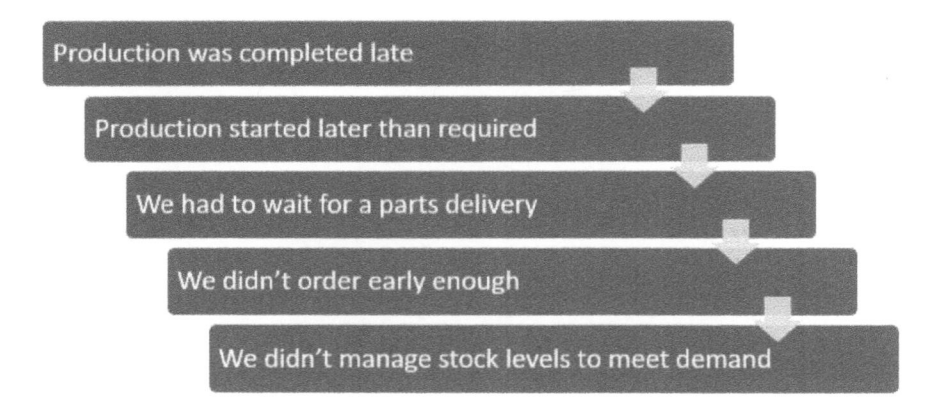

- Production was completed late
- Production started later than required
- We had to wait for a parts delivery
- We didn't order early enough
- We didn't manage stock levels to meet demand

In his book *Blackbox Thinking* (Syed 2016), Matthew Syed details how accident investigators in the Airline Industry are obsessed with methodically getting to the bottom of problems. It may seem obvious but reaching superficial conclusions will cost lives. Most notably, the Boeing 737MAX suffered a second catastrophic accident only six months after the first when data about the first crash was not sufficiently investigated or reviewed against the original air-worthiness certification (which itself was open to question).

Air crash investigators do a very deep version of the '5 Whys' method, often finding that something small is causing a problem which can be fixed relatively easily. Things like clear labelling of switches or medicines may seem unimportant but can save people's lives when pilots or medics are under pressure. Syed makes the case for capturing and analysing data without attributing blame, focusing on adding to the body of knowledge on any given topic.

Write Up Your Thinking

As we have noted before, the process of committing your thoughts to paper has several effects. The first is that it allows you to clear your mind of one thought and move onto the next. The second is that it allows you to build your understanding. As one decision or thought is committed to paper (or virtual document), it solidifies and allows the next to shape. This allows for the rehearsing and testing of thinking in a sequential way. This stream of consciousness mimics how the brain works, with relatively little multitasking being truly possible. Finally, it helps develop the onward communication of the idea if this is necessary.

For clarification, writing up your thinking is not the same as writing notes in a meeting. Those notes and the stream of thoughts are driven by the meeting, not how you might think about a challenge.

It is also worth saying that this might be better done using a journal and pen. It is also the case that sometimes your thoughts will leap around a bit. You need to go with the flow and sometimes just do a bit of reviewing to pick out the more important parts. Using mind-mapping and other forms of graphic note taking can also help.

i5: Investment (or Finding Resource)

Any RS initiative will need some level of investment. Sometimes this is only an easy redirection of small amounts of effort or money. Other times you will need to allocate significant resources to make a step change. Either way this implies an opportunity (forgone) cost so it is important that the RS initiative does not flounder. If the initiative requires explicit financial investment, then some form of return will be expected. The key point is that an effective RS solution is not free.

This investment can be in several areas, but the most common are:

- your time and the impact on your private life;
- the volume and quality of people involved in the initiative;
- working environments and productivity;
- enabling technology (information and other);
- plant and equipment;
- supplier commitments;
- research and design.

Investment is usually how the first 4 i's are achieved.

1) Intention can be started almost for free, but it will depend on the organisation, meaning you might need to create a

separate dedicated set of resources to achieve the step change. In most situations this will take the form of a project team. In *Think Simple* (Segall 2016), Ken Segall documents the specific use of Skunkworks project teams, which have their roots in World War 2 and Lockheed Martin. These dedicated teams have several pros and cons, but they usually require explicit incremental investment (although it is sometimes possible to ring-fence existing investment). Either way, you are betting on Intention being important and backed up by leaders being clear about priorities.

2) Intensity implies that either one or a group of individuals can give uninterrupted focus to a problem. In my experience, this usually involves late nights, early mornings, and weekend working. When you are on a roll with something you often must keep going. (You obviously need to be careful not to burn out or alienate loved ones. We need good managers to understand that after a period of intensity, a period of rest and recuperation is required.)

The short-term cost of this intensity can sometimes be quite low, as often it is absorbed by employees going over and above their contractual working hours. However, investment will be required to both retain your good people and keep them motivated. This can take the form of temporary or permanent salary increments, retention, performance bonuses, and even stock options. And do not forget that even non-monetary rewards like training and coaching still need investment.

3) Iterations are seemingly costly but necessary. You may have launched a product, have some initial success and now you need to improve that product.

In the world of software (including Apps) this is an established model and companies tend to invest in upgrades over the product life cycle. These have relatively low costs with limited fixed asset requirements. For certain high-tech hardware (including microprocessors), there are often very short product life cycles of nine months or a year. This requires significant thought otherwise the refresh costs would be prohibitive.

In lower margin, higher volume consumer goods, the problem is even more acute, where the simple payback is usually between three and four years. Clearly the best investment scenario is to set up a production line and let it run for ten years until the technology is pretty much obsolete and the payback and profit will have been delivered.

4) Intelligence is usually considered as an employee cost. The truth is that the more you pay, the better the capabilities, skills, or experiences that an individual (or group of individuals) will bring. As an alternative, the intelligence might come from suppliers or other sources of thought leadership, such as Universities. Other stimuli will come from any potential data-insights, so investment in information technology and data may be a key enabler.

All these investments will have a great payback, but they need to be affordable. This is why iterations are so important. You need to start generating initial benefit to fuel (fund) the next 'virtuous circle' iteration.

The 5i's Template

This RS 5i's template provides a way to assess your mindset and approach simplicity or complication challenges. If you can answer the five questions you will begin to understand the enablers of success and the areas where you may come unstuck.

REALLY SIMPLE 5i's TEMPLATE	
Date:	**Topic:**
Intention	Why is this important? • • •
Intensity	How will we maximise focus? • • •
Iterations	What are the likely iterations? • • •
Intelligence	What is the smart thinking? • • •
Investment	What resources will be necessary? • • •

Really Simple

6

Method - It's Systematic

Make haste slowly

"The journey is the reward"
-
Chinese Proverb

With an understanding of the dimensions and the required mindset, you have the foundations for tackling a Really Simple challenge. The final piece of the execution puzzle is how to do it!

Most people use some aspect of method for just about everything. In many instances it will be so ingrained that it requires little thought, although it will usually involve some teaching. For example, I was taught by a chef how to use a knife and now I never think about how I dice an onion because I have the method ingrained in me.

For more complicated and repeating challenges, it is helpful to have a more formalised and structured method, although you should never be a slave to it. You also need to make sure that the method does not outweigh the focus on the outcome; meaning the method is not an end in itself.

This chapter details the *core* method, DASIA, made up of Definition, Analysis, Synthesis, Implementation, and Affirmation. This is designed to manage an RS challenge over a life cycle of change. It will generally be used for explicit initiatives, be they a specific project or something more long-term in nature.

I also introduce 4 supporting tools:

1. I-Charter
2. DAN-RICO?
3. A-List
4. EA-List

These tools are themselves simple and support the process of creating RS solutions.

You can find these detailed in the 'Tools' section following this chapter and the complementary Microsoft Templates are available for download via the www.really-simple.com website.

DASIA

DASIA is a framework for simplification that provides the stages for a Really Simple initiative. It can be used for small and large projects and organisations.

Really Simple

Most readers will have encountered some form of project management or problem-solving methodology and this approach seeks to provide a generic Really Simple structure that can apply to all projects.

In summary, the stages of DASIA are:

- Define - Clarify the challenge.
- Analyse - Necessary understanding of the challenge.
- Synthesise - Designing the solution with simplicity.
- Implement - Putting the solution in place.
- Affirm - Reviewing and confirming that the solution met the challenge.

It is important to tailor the approach, remembering that the method is useful in a range of instances. From something individual, taking one hour to scribble on a single piece of A4, to large multi-disciplinary projects lasting many months, with many participants and tracker spreadsheets!

For those readers with exposure to Six Sigma, this will feel remarkably similar. Indeed, the genesis of this method was the DMAIC (Define, Measure, Analyse, Improve, Control) and SIPOC (Suppliers, Inputs, Process, Outputs, Customers) tools. However, their strong statistical orientation can often unnecessarily paralyse a project manager and participants. Obviously, data and its analysis are important, but DASIA is not as prescriptive as Six Sigma. Six Sigma is ideal for high volume and high repetition processes, and DASIA for more variable, RS initiatives.

If you haven't come across either of those methods before, then do not worry, DASIA is a simplified framework that can be used without needing any other background information. I have synthesised all the relevant information into a RS tool, which is easy to use and apply to everyday life, work and initiative management.

Although the level of method described here is designed for challenges that require significant resource, thinking and change, the basic structure can be used in different situations. For example, if you decide to reorganise a store cupboard you still need to go through the various stages, albeit with a bit less formality and rigour.

> It is worth repeating that the stages detailed here are necessarily quite generic, so you can use them for inspiration and develop, adapt, or personalise as you wish.

Define: Clarify the Challenge

Almost all problem-solving techniques start with getting clear on the basics of what you are tackling. RS refers to this the 'Define-Clarify' stage, distinguishing it slightly from problem solving. To ensure that this is completed thoroughly, some formality is required. The act of committing words to a document seems to magically achieve this.

Most will be familiar with either formal project charters or setting objectives. The purpose of these relatively short documents (ideally one page) is mainly to get agreement on the why, what, and how you will tackle a challenge. The key point is extreme clarity. Really Simple requires things to be really clear, and the 'Tools' section provides an *initiative charter* template (**I-Charter**) with nine 'fields'.

Creating a charter is usually put in the hands of one person but, depending on the scale, there may be several stakeholders. There also needs to be an approval method for committing resources. This might be a stakeholder meeting but if not, some formal record is usually required.

A charter will define the overall scope of an RS initiative but may not capture all the thinking. In this situation the RED-CAR model should be used in conjunction with a charter. This can be done in a few ways: either an 'as-is' assessment, or a 'to-be' assessment, or possibly both. The RED-CAR can then be iterated through the 'Analyse' and 'Synthesise' stages.

Analyse: Develop Understanding

When the definition phase has been completed, you will have an approach to the challenge, be it a problem, an opportunity, a project, or ongoing activity. The 'charter' can be used as a basis for approval to dive deeper into the necessary details. This is rarely a 'clean' process where you will leap to a good appreciation of the contributing components, so will often need more than one cycle to grasp the scope of the initiative.

There are three main Really Simple tools that support analysis, which create as much clarity as possible:

1. RED-CAR - To ensure challenges against all the major dimensions of simplicity.
2. DAN-RICO? - To capture the stream of inputs coming from all the various sources.
3. EA-List - To document any multi-step process challenges.

The main purpose of these three Really Simple supporting tools is to 'harden' the edges of the analysis, encouraging as much clarity as possible. We have already covered RED-CAR, so here I will provide a quick introduction to the purpose and usage of DAN-RICO? and EA-Lists, with more detail available in the Tools section.

DAN-RICO? is an acronym for:

- Decisions
- Actions
- Notes
- Risks
- Issues
- Challenges
- Opportunities
- Questions

It is an ongoing register of the various topics that may be important in any initiative.

Commonly a DAN-RICO? log is a spreadsheet, which can be shared amongst team members. Often it will be used in meetings as part of a review of progress and is particularly helpful in tele or video meetings. Prompting updates (ideally by the owners of the items) ahead of any review meeting is good practice.

It can also be used in a workshop scenario. It can be useful to have a wall covered with flip chart paper to capture all the DAN-RICO? insights as they come up. It takes a little practice to identify all the various types as you go along, but it necessitates agreement with the participants on the nature of the point they are raising, making it easier to address. It also means that there are no bad ideas captured, as forcing clarity on a point usually filters these out.

> Workshops are a useful way of getting a cross-section of inputs from various subject matter experts. Sessions are rarely without tension, as there will potentially be 'winners' and 'losers', therefore it is important for the facilitator to think through a few outcome scenarios. Obviously, in an ideal world, everyone will be aligned but that will rarely be an objective in itself.
>
> I have often seen a 'rules of the game' statement used at the beginning of workshops that asks everyone to be open and honest. This always confuses me, as if asking someone who was planning to be dishonest, not to be, would magically work. More effective tone-setting statements include:
>
> - Respectful but tough love required.
>
> - Nothing important should be left unsaid.
>
> - Stay in the room, not distracted by tech.
>
> Ideally you will have a scribe capturing the DAN-RICO? directly into the electronic log which can then be issued almost immediately after the meeting.

It is also quite common and helpful to create some architectural 'models' to help articulate something that is complicated. (Architecture in this context means a structure, and a model means a representation of reality.) Usually, these models take the form of a diagram like an organisational structure or a flow diagram. Mostly these are straightforward, but process redesign does not seem to have solution for non-experts to create, iterate, and share their thinking and conclusions.

An **EA-List** is a RS way of documenting processes using **events** and **actions**. It is a simple and robust method, as processes are documented as a series of events or actions in a table format. You also specify the actor, which is the person, role or system triggering an event or completing an action. Finally, there is a section to cover clarifications which also uses the DAN-RICO? labels.

Using an EA-List effectively takes a bit of familiarisation time. Understanding events and decomposing actions to the right level of detail is not a mathematical equation, it takes judgement. An event can be triggered by a moment in time (e.g., 12 p.m. daily), a status changing, something being completed (e.g., a form), or something external (e.g., receiving an email). An event can also include problems and encourages the user to think through both the happy and unhappy path through the process.

The strength of the system is that it can be understood universally and is easy to walk through. In particular, it encourages iterations by encouraging clarifications to be captured as questions that need answering.

Synthesis: Designing RS Solutions

At some point during the analysis phase, the brain starts to wander towards solutions. I do not discourage this (as I mentioned previously, it's great to capture your first instinctive thoughts), however, I would also advise you to proceed with caution. This is because ideas based on limited or undeveloped analysis are often, in truth, no more than personal hunches. Personal hunches are not a bad thing, but at this stage they simply need to be recognised, particularly with people who have successful track records but are now applying themselves to a new challenge.

Experimentation

It is important to understand that it is unlikely (almost impossible) to leap directly to the right solution. To some extent, experimentation is the hard work of design because there might be some blind alleys and some seemingly wasted work. In traditional corporate life, this is sometimes discouraged as a waste of time. Clearly you cannot spend too much time experimenting, otherwise you never quite reach a conclusion.

Prototypes

Ideally, you will be able to pick one or two of your RS solutions and build some form of prototype to test if it is implementable.

Implementable?

A key part of synthesis is establishing the potential for practical implementation.

Here are some key questions to consider:
- **Legally** Are there any legal issues?
- **Procurement** Is there a supply chain challenge, remembering that fewer dependencies are better?

- **Capabilities** Do you have the right technologies and resources?
- **Manufacture** Can you make the volumes required at the right quality?
- **Economics** Is it cost-effective and value-generating?
- **Schedule** Is there an achievable schedule given the resources and any other constraints?
- **Risks** What are the main implementation risks that are not addressed?

Implement - Change/Create

An implementation will usually comprise either the changing or creating of something. A great Really Simple design is not enough by itself, you need to be able to convert it into something real.

To do this you need to make sure that you have good plans (activities, resources, and timing) that allows for preparation, cut-over, and initial support. There are a wide variety of ways to create a plan but if it includes significant amounts of people and resources, an A-List (Activity List) can help.

An **A-List** provides an RS way of capturing and sharing expected activity (the plan/estimates), tracking actual progress, and refining the estimates. The planning and tracking is bottom-up, so they rely on those allocated with the capabilities and skills to create their own estimates and measure their own progress.

Usually an A-List will be held in some form of spreadsheet or table, which can provide analysis summaries of progress. You can create columns to categorise important types of analysis such as priority or resource. The primary benefit is the clear capture of the specific resources that need to be used and when.

Preparation

Create Capability - This can range from the very simple ordering of equipment to the very complex building of a production line or hiring/developing new skills. (The most common in business will be building/changing a process or software, or launching a product.)

Outcome Trials (Testing) - Commonly the outcome you are working towards will be new and will need some sort of trial or testing.

Change Management - Oftentimes some form of education or training will be required. Getting this right is relatively difficult as often the actual change is taking the lion-share of the team's energy, so making sure that this gets dedicated or enough resource is critical.

Cutover Rehearsal - If the cutover has many moving and interconnected parts it is important to do a run through a few times. This will shake out the gremlins and confirm the estimates of the time that the various steps will take. It is common for people to feel that they can do this all in their heads. My experience is that the opposite is usually true, and without rehearsal they will either forget necessary tasks, underestimate the time required to do those tasks or something predictable is missed.

Cutover

Usually, there is a moment when either the change or the new solution is live. This may mean stopping something that is currently being used or doing some one-off activities like a data load or a backup. It will require an organised schedule and good communication between all those involved.

Support

Just after go-live, some extra support is usually sensible to mop up any issues that have not been caught through the project. This is usually measured in days or weeks and sometimes involves having suppliers being on standby or in attendance.

Affirm: Review

The final, and often overlooked stage, is to affirm that the objective was achieved as per the original charter. There should be a final close down of the initiative, even if it's to authorise a party! Sometimes the affirmation is about recognising the learning that will help in the next iteration.

DASIA Template

Depending on the size of your challenge, you might choose to use any of the supporting tools. In many circumstances it is sensible to use this DASIA template to formally document your thinking and approach (to the steps). It is probably only a ten-minute exercise in many situations, but has a checklist effect, requiring you to think holistically and a few steps ahead.

REALLY SIMPLE DASIA TEMPLATE	
Date:	**Topic:**
Definition	How we will • • •
Analysis	How we will • • •
Synthesis	How we will • • •
Implementation	How we will • • •
Affirmation	How we will • • •

Tools - Supporting

Whilst RS is about understanding dimensions, psychology, and the need for a systematic approach, we have introduced a few tools that help tackle the most common problems. In addition, there are some other RS tools that might be useful in more specific contexts which are outlined in this section.

The first two are focused on the individual and can be used on a personal level, whilst the third and fourth will support any initiative. And five and six are used in an enterprise setting.

RS Toolbox	Personal 1. DF-Plan 2. S-OPRA	Initiative 3. I-Charter 4. DAN-RICO?	Enterprise 5. A-List 6. EA-List

These descriptions include images of potential templates, and you can find more on the Really Simple website (www.really-simple.com) or build your own.

If you want to go a step further and use these tools as part of a systematic approach, there is a set of templates built in Coda, that provide further efficiency and rigour. Enterprise options are also available but are usually bespoke to the situation.

Tool 1: DF-Plan

There are hundreds of different ways of managing your time and action management, and it would be helpful if we could just pick one. However, the reality is that we all live our lives in slightly different ways and the tools and expectations will be varied.

The DF-Plan focuses on planning and managing your daily flow (DF) of work. This system is focused on minimising the lifting and laying of the various inputs in the system, preventing items from dropping through the cracks. It will not be right for everyone, so you are encouraged to adapt it to suit your needs.

Note. This system is not designed to define what you should be working on, but rather once you have decided, making sure that you follow through on it.

The basic concept is to think about the daily source, flow, and dependencies of any activities you need to address. The focus here is on making sure that you can get the important work done with maximum efficiency, by creating a strong understanding of your daily flow of work.

This system recognises that you will probably have activities in a backlog of outstanding items that you need to track and categorise. This is called the Categorised Master List (CML). In addition to this, you will likely have significant activities driven by email and meetings in your diary. Taking these three inputs you can create your daily plan.

Clifford Burroughs

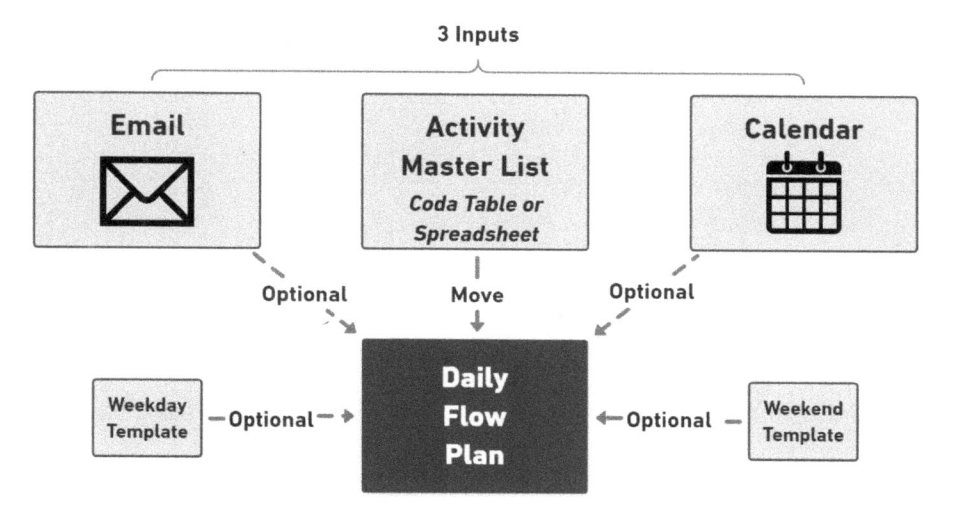

3 Inputs

```
   Email          Activity Master List          Calendar
                  Coda Table or
                  Spreadsheet
      \                |                    /
       Optional       Move          Optional
            ↘          ↓          ↙
  Weekday                                    Weekend
  Template  —Optional→  Daily Flow Plan  ←Optional—  Template
```

The key to the system is categorisation that is logical, quick, and simple. If it is not logical and quick, it will become more effort than it is worth. For example, you will need a way of identifying which emails you need to act on, or where you are waiting for a response to take the next step. Some of the actions are not your own but you will still want to monitor progress.

Also, not all activities are certain, so there needs to be categories for 'maybe' and 'someday'. These are great for those things that you want to note down so you will not forget them. This might include social things like holidays and seeing friends, or longer-range thinking about initiatives.

Really Simple

Categories might include:

Stage	Nature	Initiatives
Action Waiting Monitoring Decide Someday Maybe Scheduled	Work Chores Leisure Hobby Exercise Social Community Learning Networking	Projects Ongoing

Each of these categories can have subcategories. For instance, in exercise I have sub-categories of swim, cycle, walk and row.

The projects and ongoing actions need to be categorised specifically. For example, in my case, I have a project called 'Book' with sub-projects that include, content, publishing, and promotion. For this large project, I just have a high-level action in the DF-Plan, like 'Book-Content Review Chapter 4'. I then also have a separate and more detailed plan. This categorisation helps clarify my actions but also highlights when I am starting something that is not on my 'expected' list.

Much of this approach and terminology was developed by David Allen as part of his *Getting Things Done* (Allen 2015) (GTD) method. There are lots of ways in which you can implement this system, but the most important thing is to consciously analyse the nature of your activity demand. This analysis needs to encompass your work and private life because of the amount of interplay between the two.

Tool 2: S-OPRA

Most of us need to store and access many different types of information in both our work and private lives. The way in which you store this information has the potential to drive easy access with minimal mental friction. The fundamental challenge to easy access is easy filing. If you do not have a method that makes filing easy, your system will quickly break down under the weight of other pressures.

OPRA is the macro taxonomy (a way of organising) 'storage' structure, comprising ongoing projects, references, and archives. This is augmented with a 'staging' section, for things that are still being processed.

This structure is predicated on the belief that you have to manage what is 'coming at you'. Some will be self-created, and some will just land at your door, either way, a conscious process of inbound management will give you a better sense of control.

There is an argument that with new search tools, tagging or filing is no longer necessary. There is an element of truth in this, but it depends on embedded keywords that match your mental map. I often search using a keyword that is obvious to me but remain frustrated until I pick the right term that matches what I am searching for.

The examples below are illustrated using Gmail and Google Drive. Other collaboration tools will have the same types of options. Therefore, the solution provided here is quite general and transportable to the other technology platforms.

Really Simple

This tool is designed on the premise that proactive organisation of data is important. That in turn recognises that without some conscious system you almost would not know what you can depend on; a bit like searching the internet, you may or may not get what you are looking for. Also, most organisations have established expectations of the materials and documentation that support management and therefore need to be properly accessible.

Ideally, much of this categorisation or labelling will happen automatically using your chosen technology. If not, then it needs to be quick and easy, or you probably won't do it.

The OPRA model is mainly designed to categorise documentation for the longer term. Given that the nature of some work is not at the outset entirely clear, the Really Simple S-OPRA model adds an optional 'staging area'. The risk is that this becomes a bit of a dumping ground, like an overflowing inbox.

Staging: For unfinished items, sometimes with dual labels in the ongoing or project sections.

Ongoing: For topics that are ongoing and usually regular or repeating.

Projects: For initiatives that will have defined output and will be closed at some point.

Reference: For information that you might want (interest/knowledge) or need (legal) access to in the future.

Archive: Used mainly for projects that have been completed.

Clifford Burroughs

Depending on your email and office platforms your implementation will follow the same principles. Let's take a look at each category and how it might look, using real examples.

In Google Drive at the top level, it could look like this:

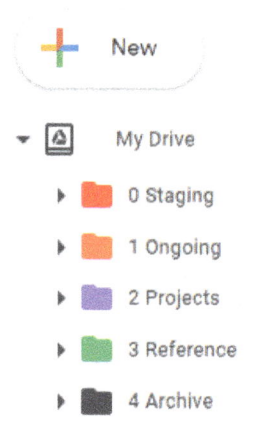

- New
- ▼ My Drive
  - ▶ 📁 0 Staging
  - ▶ 📁 1 Ongoing
  - ▶ 📁 2 Projects
  - ▶ 📁 3 Reference
  - ▶ 📁 4 Archive

In Gmail at the top level, it would look like this:

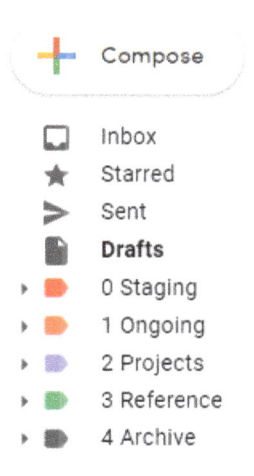

- Compose
- Inbox
- Starred
- Sent
- **Drafts**
- ▶ 0 Staging
- ▶ 1 Ongoing
- ▶ 2 Projects
- ▶ 3 Reference
- ▶ 4 Archive

Staging

Staging will be specific to the type of email you receive and how you process it.

My staging area is structured as follows:

- 0 Staging
    - Action
    - Decide
    - Delete 28
    - Delete 365
    - Waiting

Label	Explanation
Action	I need to do something with this email.
Decide	There is a decision about what to do next.
Delete 28	I almost certainly don't need this but will keep for 28 days just-in-case.
Delete 365	I almost certainly don't need this but will keep for one year just-in-case.
Waiting	When I have sent an email that I am waiting for a response on.

Ongoing

In most situations, below the top level, alphabetical filing is the most sensible. For 'Ongoing' it may be helpful to further subdivide according to periodicity, either years, months, weeks, or whatever is appropriate. For example:

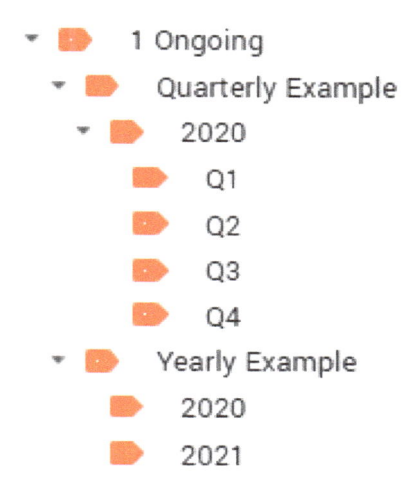

- 1 Ongoing
    - Quarterly Example
        - 2020
            - Q1
            - Q2
            - Q3
            - Q4
    - Yearly Example
        - 2020
        - 2021

Projects

For projects they should just be labelled alphabetically (shown here as A, B, C). As projects finish, they should be moved to the archive to prevent the list getting unwieldy.

- 2 Projects
    - A Project
    - B Project
    - C Project

Reference

Reference should also be labelled alphabetically. This list can get quite long, therefore some of the reference does, over time, need to go into the archive folder.

- 3.Reference
    - A Topic
    - B Topic
    - C Topic

Here are some example categories that I have used both in Gmail and Drive. These are relatively generic and can be mixed with more specific topics like 'Microsoft' or 'Hilton'. You can determine the level of specificity you want to go into. For example, I break 'Network' into three levels (high, medium, and low, not shown here) to reflect the importance of the relationship.

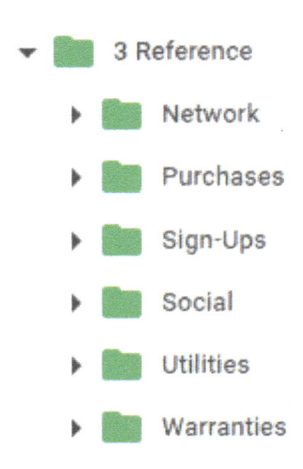

- 3 Reference
    - Network
    - Purchases
    - Sign-Ups
    - Social
    - Utilities
    - Warranties

Archive

The archive folder will be a mix of everything that you want to store long-term:

- 4 Archive
    - D Ongoing
    - E Project
    - F Topic

Eventually you may need to delete things from your archive and you will need a bespoke solution for deciding this, although age is often the key criteria.

Implementation

Depending on the situation you might need to divide the folders by work and personal, but most of us have two email addresses to manage this separation.

The model can also work for teams if they have shared folders or inboxes.

Most email programmes can support labels, filtering, colours, and auto-labelling functionality. The most important thing is to have a systematic approach to both decide the right folder and then quickly file without requiring much thought.

If your email inbox and folders are overloaded and a bit of a mess, you will find it quite hard to tidy it up unless you lock yourself away for a week. The best solution to this is to:

Really Simple

1. Setup the folder structure as above.
2. If you have some elements of reasonable organisation, migrate that to the OPRA model.
3. Do a bit of bulk 'housekeeping' where you delete all the obvious rubbish by sorting and filtering.
4. Setup auto-labelling of all the key ongoing and project topics.
    a. This will help you identify between the two types.
    b. It will help tackle the backlog in your inbox.
5. Create a 'historic' folder in archive and move everything in your inbox that is more than one month old to that folder i.e., don't delete it.
6. Process the remaining inbox emails from the last month into the right folders.

This may still take you up to a day, but you will have broadly reset your ability to process your inbound messages, and not have invested too much time categorising old emails.

You obviously need to be really careful when doing these mass clear ups. It is easy to press the wrong button and delete the wrong thing, sometimes en masse. If possible, take some form of backup before you start and ideally be purposeful and concentrated as you delete or move things. You will already have a good basic idea of how your software works, but make sure you understand the various functional options and features before starting.

Tool 3: I-Charter

An I-Charter is for initiatives. Usually, they are one-page documents detailing a project, but they can also detail the requirement to make an ongoing improvement. The main purpose is to get a formal agreement and buy-in on the approach and resources necessary to tackle the challenge.

Normally, one person is in control of creating the document, but there may be a number of stakeholders who need to contribute to or approve the approach. This might be as simple as meeting, but sometimes a formal record is helpful, particularly where there are significant risks or resources required.

Really Simple

An I-Charter should comprise the following key attributes:

Attributes	Description	Example
Objective	This is the overall aim of the project and may be reasonably general given that other attributes will provide more specific details.	Grow healthy snacking offering by developing sugar replacements.
Scope	It is useful to qualify scope by clarifying what is 'in-scope' and more importantly 'out-of-scope', as this may be where expectations are confirmed.	Focus on a few core offerings. Focus on UK Market Solutions. Calorific value is not a major target. Focus is on product, not marketing.
Outcomes	This is a physical outcome without volumetric detail. For example, a facility, product, or process.	Three new products with healthier credentials that don't cannibalise existing revenues. with margins maintained or improved.
Timing	An overview of the key timing or milestones. The detail depends on the scale of the project, meaning sometimes it will be in weeks, months, quarters, or even years.	Start Mid 2021. Complete by end of 2021.
Resource	People or assets that are required for this project to be delivered.	Core R&D team need to provide three FTEs for six months. Marketing and Testing Resource required. Trial capacity needed every month for a 0.5 shift.

Clifford Burroughs

Measures	Some specific KPIs that make sense in the context of the initiative.	Incremental £1.5m revenue and 0.3m profit.
Risks	Thinking about what might go wrong. As the project progresses there may be more detailed risk analysis, but you should highlight likely themes here.	Consumer testing is negative. Production is difficult with existing equipment. Cannibalisation of existing products. Limited retailer interest.
Constraints	These are the things that will slow you down or get in the way of likely success. Documenting these will either cement acceptance or encourage solutions.	Initial short manufacturing runs, will test manufacturing costs.
Recommendation	Sometimes a specific recommendation is required.	Proceed but with regular stop-go assessments. Hold for now.

Really Simple

REALLY SIMPLE **I-CHARTER** TEMPLATE		
Title:	**Date:**	**Owner:**
Objective	Timing	Risks
•	•	•
•	•	•
•	•	•
Scope	Resource	Constraints
•	•	•
•	•	•
•	•	•
Outcomes	Measures	Recommendation
•	•	•
•	•	•
•	•	•

A Really Simple (**I-Charter**) template is available for you to use on www.really-simple.com.

Tool 4: DAN-RICO?

Over the years I have seen my fair share of trackers. One of my ex-colleagues once quipped that if business success was measured in trackers, we would have been the market leader. The reason they are so important is that the brain is not good at keeping track of many things at once. That is why, since the beginning of time, mankind has used all sorts of techniques to not forget the things that are important.

I noticed that one of the complications, particularly in significant projects was a proliferation of trackers, sometimes overlapping each other, but rarely was there one version of the truth. Team members also often kept 'off-line' personal trackers impacting the visibility of the potential workload.

I also realised that the trackers were often a reflection of the author, rather than an accurate and concise record of what the project was doing. I looked for ways to 'harden the edges' and simplify the process, so I developed DAN-RICO?.

The basic premise is that there should be a log of all the important topics. Having worked on lots of projects and having seen lots of trackers, I realised that there were reoccurring types of topics.

DAN-RICO? is a tracking tool to provide a unified log for all relevant topics. DAN-RICO? stands for Decisions, Actions, Notes, Risks, Issues, Challenges, Opportunities, Questions. A log structured in the right way could be filtered by either:

- the person who logged the entry;
- the person meant to be actioning the entry;

Really Simple

- the importance/urgency;
- the theme/ or workstream.

Initial versions of this were held in a spreadsheet but latterly I have used Smartsheet and now Coda. These cloud-based tools allow easy access across teams and divisions and provide a few critical workflow automation features to encourage everyone to keep them up to date. The very best way to ensure that the list was kept current, and that entries were being tackled was a weekly review with the key stakeholders. Easy completion meant that the document could be updated live in the meeting.

One of my other observations was that items that seem important one week might magically solve themselves and not be a problem a few weeks later. The very act of recording something clarifies understanding and can almost be imperceptibly virtuous.

Whilst the template is fairly self-explanatory, I do sometimes get asked when to use the various elements, so I have provided the following breakdown.

Attributes	Description
Decision	Often decisions get taken on a project, then someone records them in an email or minutes, and you may be at the mercy of people's memory. The D of DAN-RICO allows you to run a log of those decisions and look at them as and when necessary. Often a decision is the result of the solution being decided for one of the other items. For example, you might have had a question (Q) about the right time to launch something, but when you decide the answer, the log entry is converted to a 'D' and marked as closed.

Action	Not to be used for every action, but just those that are critical to the collective success of the initiative. For example, you may be relaunching a website so an action to identify content owners may be required, but you wouldn't record each element of the content that needed an owner.
Note	This is used to record a useful fact or a clarifying point. In a systems context, a good example would be the number of people using a system. This can also be used for references and links to other useful information.
Risk	This is something that might happen, indeed might even be likely to happen, but has not yet happened. An obvious and common example is a performance issue on a system at the point of go-live. Another is suffering a cyber-attack.
Issue	An issue is something that has already happened and might have previously been an identified risk. The most common example is loss of critical resources for one reason or another.
Challenge	A challenge is different from an opportunity, in that an opportunity does not necessarily recognise any constraints and, therefore, may not be included in the solution. A challenge, therefore, is something that should be tackled within the project, which is necessary to achieve the objectives.
Opportunity	Opportunities can be used in two ways. Firstly, in the classic marketing opportunity sense, but more commonly it is used to log ideas or improvements that could be included, perhaps not now but in the future.
Questions	This is often the most important element to capture as it provides a holding place for the known unknowns. At the end of the project there should be no open questions, as all should have been answered, closed, or converted to one of the other categories.

There are two main scenarios to utilise DAN-RICO?: workshops and initiatives.

1) Workshop

Often in a workshop, there is a stream of thoughts being shared, with some level being captured by the facilitator or the scribe. A simple way to collect these is to recognise whether a point is one of the DAN-RICO? components. If it is, scribble the letter and the point on a flip chart, then put the various flip charts around the walls. This makes the process frictionless, interactive, and very efficient as you agree which DAN-RICO? element is right.

> When I am likely to use this technique, I check whether I have enough wall space for the number of flip chart pages I might use. I also have someone typing up the points into the template, so I can share the output quickly and succinctly as part of a workshop output pack.

2) Initiative

In most projects there will be some form of status log and DAN-RICO? should simply replace those. These should be shared across the various team members, usually workstream leads, and they should be encouraged to directly input their entries. They then need to be continually reviewed by the project manager.

I am sometimes asked whether everything should go in the list, and the answer is definitely NO! The suggested criteria for items going on the list are:

a) items that are too complex to for a simple email exchange;

b) when it needs to be proactively shared and understood;

c) when the likely solution is ambiguous or very problematic for the project.

You can use the log how you want but the critical success factor is its regular use.

> I also use these DAN-RICO? categories for personal notes that I make, and also in the EA-List tool.

A Really Simple DAN-RICO? template is available for you to download on www.really-simple.com.

Tool 5: A-List

Despite a plethora of project management thinking, practices and tools, good project delivery remains elusive. This is particularly true where the significant resource is knowledge work, combined with a subjective view of success. Whilst I have seen projects like this deliver, it is in spite of all the planning and controls and more through the heroics of the team. I have also seen projects deliver that have had overwhelming levels of resources but often even these seem a struggle.

So rarely do projects go according to plan, being roughly on-time, reasonably on-cost, and meeting expectations. There are obvious reasons for this; the unknowns that come from nowhere, the optimism bias when constructing the plan, and a misunderstanding of the true nature of the objective. Over the years, I have worked to unearth the less obvious reasons and found that in the end they were simple.

I came across a single line of wisdom that unlocked my thinking. This was in the book *Rapid Development* (McConnell 1996) by Steve McConnell. At the time I was working in software development and was involved in several projects which felt like death marches. Some of these projects already had a limited chance of success, but by putting in 70 hours a week, you could get them done. I then went for an interview for a trendily titled Rapid Application Development (RAD) role. The person interviewing me actually described the need to do 100-hour weeks, not continually, but still, I knew there had to be a better way. So, I found Steve McConnell's book, which made the case that true speed (as opposed to perceived speed) was not dependent on any new-fangled technique, but more on first principles. Over a few pages of pure gold (pg 46), he summarises 36 classic mistakes, and whilst all are important and fairly intuitive, number 25 was "Omitting necessary tasks from estimates." He then made the point that this single mistake might add up to between 20 and 30 percent of the total required.

This was a lightbulb moment for me, as I had example after example of where the core activity was fairly well planned but other peripheral, yet critical, activity was not.

A simple example is the need for training organisation. I have seen plans that include the development of training and delivery of training, but not the organisation of the training. The organisation, whilst a seemingly trivial task, falls on someone in the project team but eats buckets of time across all the various stakeholders. The overall time required for training is higher than the plan captured, but the omission is missed, and in the heat of battle everyone soldiers on. There is always a price paid for this miss, either in quality, time, or cost.

The painful truth is that many project plans are often not much more than a good start. This is because individuals or their managers are rarely good estimators of effort and there is also a tendency whereby required tasks are completely missed from the plan. Recognising this reality changes the way you plan, the way you recalibrate those plans, and the management of the overall resource available.

The reality is that you need to understand and methodically capture and update all tasks. Most importantly, you really

Really Simple

need the people that are doing the work to **own the accuracy of those estimates,** and ensure that they understand the true effort required for a task.

Obviously, this is not the only thing that you need to get right to have a successful work project. However, these Really Simple templates can provide the core of a project management system, which can be supplemented by some of the other tools you consider important.

The core principle is that you capture every activity required to complete a project, by individual, at the day level. If something will take a significant chunk of a day to complete, it should be captured. The smallest unit I use is a quarter of a day, but ideally items will be captured at the half or full day level. These activities are then allocated to a week (or a range of a few weeks). Here is an example:

Work Type	Description	Priority	Resource	Orig Est Work Time	Act Work Time	Lat Est Work Time	Work Time Diff	Orig Est Start	Orig Est Comp	Act Start	Act Comp
Training Des	Using Coda	B	CBU	1	2.25		1.25	12.10.20	16.10.20	13.10.20	16.10.20
Training Dev	Using Coda	B	CBU	1	0.75		-0.25	19.10.20	23.10.20	20.10.20	21.10.20
Training Org	Using Coda	B	CBU	1	0.25	1	0.25	26.10.20	30.10.20	26.10.20	
Training Del	Using Coda	B	CBU	1			0	2.11.20	6.11.20		

This shows four tasks, differentiated by their work type, allocated to me, and originally estimated to take a day each. After a few weeks, the work has been completed on two tasks and the organisation task is still in flight, with the final task having not been started. This list shows a mixture of slight over-run (in amber), a higher over-run (in red) and an under-run (in green).

On a weekly basis, every team member will review their individual plans by filtering on themselves and capturing the actual time worked on a task as accurately as possible. They will then re-calibrate the work required to complete any outstanding tasks, including the work already done. This also gives an opportunity to add any missing tasks and, by implication, flag this to the project manager. Most importantly, it allows visibility of the workload by individual and by week, therefore, stopping unsustainable overloading occurring. (In sophisticated versions of this set of templates you can even allow for individual holidays.)

The model also encourages the concept of prioritisation, meaning that often something that was important at the start of a project, might fade away as time goes on (or indeed vice versa). You can model this either by using the 'Priority' column or rescheduling tasks, rather than deleting the item, ensuring that you have a running backlog of potential tasks. (This also means that you could repurpose the solution to work in focused 'sprints'.)

Priority Examples:

A – Critical

B – Important

C – Useful

H - Held

Really Simple

The A-List provides some intrinsic reporting through simple filtering and formatting rules; however, you can develop numerous roll-ups and crosstabs that provide insight. Some of that reporting might require snapshots and need to be presented in slides or spreadsheets.

The critical success factor is the regularity of the update. Friday afternoons or Monday mornings tend to work best for the cut-off. The habit of review will mean that over time the quality of estimation and planning will improve, and that project management has true visibility of status.

Whilst initially this seems a lot of detail, an individual will have between 10 and 100 tasks to review weekly and most of them will not be updated at the same time, so it is far less onerous than it looks. Once you are used to the system you might wonder how you survived without it, because it brings a level of clarity, ownership, and structured thought to your workload.

A Really Simple A-List template is available for you to download on www.really-simple.com.

Tool 6: EA-List

Process mapping is an important element in many businesses. It provides a way of sharing understanding about the current process and sometimes the future design. Having been involved in several big systems re-engineering projects, it was clear that often after the project these process maps had a few problems:

- The method of recording the ancillary details that related to each step of the process was not robust. All the various mapping software has different methods for doing this and to some extent this also causes a problem.
- The physical layout of process diagrams on the page takes a disproportionate time, especially as new elements get clarified along the way.
- Process walk-throughs were often hit and miss because they were rarely sequential and difficult to do electronically.
- Keeping testing and training documentation for users 'in-sync' with process changes was highly dependent on the diligence of the individual involved.
- Process documentation is often skewed towards detailing a 'happy path', where all the right steps happen, as opposed to what really happens.
- Training and use of the process mapping tools (including Visio) was extremely varied. Also, not everyone had a licence for these flow diagramming tools (because they are expensive).

This meant that they often fell into disuse, especially when the person that created the document moved on to a new assignment.

So, I developed the EA-List method, which has been used repeatedly and successfully, with the main challenge being that it almost seemed too simple until people got the hang of it. Originally it used Excel, which is still a good solution (a template is included below). However, the more advanced Coda Doc versions transform the efficiency of keeping the various documents up-to-date and in-sync.

The core premise of the EA-List is that every process can be broken down into a list of **Events** that start one or more **Actions**. These are listed out in a table, sequentially, without any diagramming or shape-linking required. The order of the events and actions is usually important and shown simply by the order they appear on the list relative to each other.

A process always starts with an event. The first action must be preceded with an event, but subsequent actions can chain together.

E/A	Description	Actor	Clarifications (DAN-RICO?)
E	Want a cup of coffee	Cust	N Assumes you are away from home.
A	Find an open coffee shop	Cust	Q Are there any other decision points, like which brand or length of wait?

Usually, the actions are obvious, but the events are less so.

The Actor column provides a short-form description of the thing or role that is facilitating the event or action. (In diagramming methodologies this is often achieved using 'swim-lanes', which cause layout headaches that this approach avoids.) In the example, the roles are a Cust(omer) and SA (Shop Assistant).

The 'Clarifications' column is critical. When documenting a process, it facilitates the capture of all the questions, risks, issues, or challenges you might have. As you answer those questions, or get clarity about the risks, issues, or challenges, you will be getting closer to an 'as-is' or 'to-be' final version.

The simple example below provides a more complete overview of the process of getting a coffee from a coffee shop.

E/A	Description	Actor	Clarifications (DAN-RICO?)
E	Want a cup of coffee	Cust	N Assumes you are away from home.
A	Find an open coffee shop	Cust	Q Are there any other decision points, like which brand or length of wait?
E	Found an open coffee shop	Cust	
A	Get in queue	Cust	
A	Decide on exact order	Cust	
E	Shop assistant ready to take my order	SA	
A	Place Order	Cust	N Will need to respond to all the up-selling questions.

Really Simple

E	An aspect of the order cannot be fulfilled	SA	
A	Revise my order or abandon	Cust	
E	Shop assistant ready to take my payment	SA	
A	Make Payment	Cust	N Obviously there are different types of payment, and a payment card might be rejected.
E	Order Ready	Shop	
A	Grab Coffee	Cust	
A	Leave Shop	Cust	

Believe it or not, this is a simplified view of the process with lots of other events that take place subliminally, like giving your name, getting the right milk/sugar, and choosing the form of payment etc. I have not documented all that detail because it is not necessary, as it is such a familiar and common process. The important point is that using this starting framework you could add those details if they were important or useful.

This is also an example of a broadly 'happy-path' process where basically everything goes right. The only 'unhappy-path' moment is when an aspect of the order cannot be fulfilled, meaning you must revise or abandon the order. This illustrates another key point, that there is still quite a lot of subjectivity in how to document processes, which relies on the author(s) understanding the nuances of what is important and not.

Another important aspect of the system is that it forces brevity and clarity at the same time. Once you have had some practice, it is easy to type in concise, bullet form entries that clearly define what is happening. If it is done well, reading down the description column should make stand-alone sense, with the 'Clarification' points providing useful extra information.

Some might be looking at this and thinking that this is not process documentation, but rather more like a work instruction. To some extent that is true, and it is designed to be able to do both, so it is dependent on the level of detail that you choose to include. For example, you could add screenshots to an EA-List and that would transform its usage.

The example provided here is of a single process, but in reality, all organisations are a myriad of processes, usually with some sort of functional or process structure. One of the critical needs is to have the macro process structure understood so that you can break it down into meaningful chunks. For ease, this is usually hierarchical in nature but should ideally be only three layers deep.

Here are some examples of Events, Actions and Actors:

Really Simple

Attribute & Description	Examples
Event An event describes some form of trigger or state-change and is the start of one or more actions.	A mouse click on a screen The population of a field on a screen The completion of a form A specific moment e.g. the Y2K cutover A repeating moment e.g. Weekly, Monthly A delivery being delivered An order being placed or received A email being sent/received A telephone call being made/received
Action The thing that needs to be done.	The actual filling in of a form The actual creation of an order The movement of a physical thing The creation of a spreadsheet The extraction of some data The manufacture of an item
Actor Each Event or Action has an actor. The actor is the thing or the role either making the event happen or doing the action.	A specific system A specific role e.g., a Payroll Manager, a Warehouse Operator A location
Clarifications This is a way of 'qualifying' the Event or Action. For ease, it uses the same notation as the DAN-RICO? toolset highlighting significant process detail.	DAN-RICO?

A Really Simple EA-List template is available for you to download on www.really-simple.com.

Clifford Burroughs

Summary: It's Really Simple

Really Simple is a systematic approach to conquering either existing complications or preventing the creation of new ones. It uses a structured approach, attacking the specific problem at hand whilst challenging your mindset so that you see things through a simplicity lens.

This book covers a lot of ground, so this schematic diagram summarizes how the core components of Really Simple fit together.

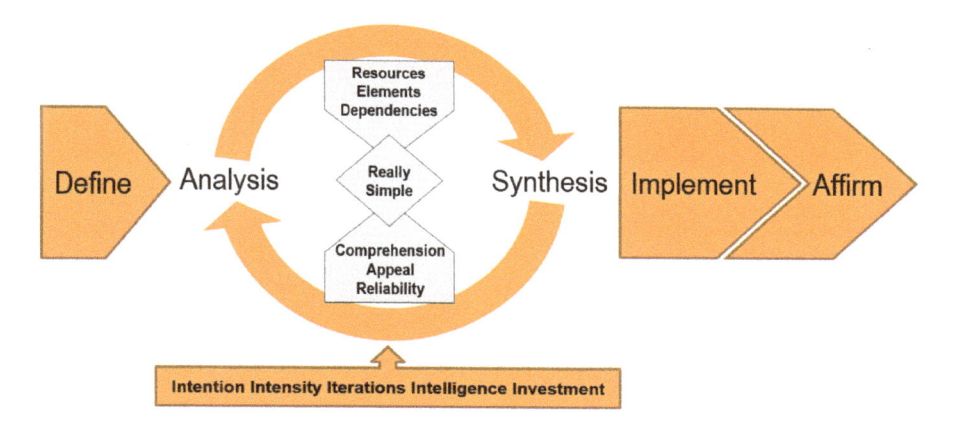

Define › Analysis → [Resources Elements Dependencies / Really Simple / Comprehension Appeal Reliability] → Synthesis › Implement › Affirm

Intention Intensity Iterations Intelligence Investment

The DASIA methodology runs over time as you move from one stage to the next and the circle represents the likely need for iterations on 'Analysis' and 'Synthesis'. At the heart of the iterations is the RED-CAR model which recognises that you need to decrease any building-blocks and increase the human connection. **You create a Really Simple solution at the balance point.**

Really Simple

The model is underpinned by the 5i's mindset. Using these 5i's ensures the DASIA method and RED-CAR model will succeed.

My aim was to explore why making things Really Simple is important, what it means in real terms, and how you can use it to simplify and improve your life and work. A Really Simple approach is not always easy and certainly not simplistic. Sometimes, it can even seem counterintuitive.

I hope that you can see how to navigate this important topic with more rigor and confidence, and that the practical tools offered give you a way to get started on your Really Simple journey.

Downloadable templates are available on the website, and we offer more advanced options, bespoke enterprise solutions, and business consultations. Visit www.really-simple.com for more information or to get in touch.

Bibliography

Allen, David. 2015. *Getting Things Done: The Art of Stress-free Productivity.* Piatkus.

Ashkenas, Ron. 2009. *Simply Effective: How to Cut Through Complexity in Your Organization and Get Things Done.*

Baumeister, Roy F., and John Tierney. 2012. *Willpower: Rediscovering Our Greatest Strength.* Penguin.

Bono's, Edward De. 2015. *Simplicity.*

Christian, Brian, and Tom Griffiths. 2017. *Algorithms to Live By: The Computer Science of Human Decisions.*

Churchill, Winston. 1948. *The Gathering Storm.* RosettaBooks.

Collins, Jim. 1975. *Good to Great: Why Some Companies Make the Leap...And Others Don't1975.* HarperBusiness.

Gladwell, Malcolm. 2005. *Blink: The Power of Thinking Without Thinking.* Penguin.

Kahnerman, Daniel. 2012. *Thinking, Fast and Slow.* Penguin.

Maeda, John. 2006. *The Laws of Simplicity (Simplicity: Design, Technology, Business, Life).*

May, Matthew E. 2012. *The Laws of Subtraction: 6 Simple Rules for Winning in the Age of Excess Everything.*

McConnell, Steve C. 1996. *Rapid Development: Taming Wild Software Schedules.* Microsoft Press.

Schwartz, Barry. 2004. *The Paradox of Choice – Why More Is Less.* Harper Perennial.

Schwartz, Tony. 2010. *The Way We're Working Isn't Working: The Four Forgotten Needs That Energize Great Performance.* Free Press.

Segall, Ken. 2015. *Insanely Simple: The Obsession That Drives Apple's Success.*

—. 2016. *Think Simple: How Smart Leaders Defeat Complexity.* Portfolio Penguin.

Syed, Matthew. 2016. *Black Box Thinking: Marginal Gains and the Secrets of High Performance.* John Murray.

Lightning Source UK Ltd.
Milton Keynes UK
UKHW020946191021
392419UK00007B/516